LF Today

a guide to success on the bands below 1MHz

Third edition

by
Mike Dennison, G3XDV

with additional material by Alan Melia, G3NYK

Radio Society of Great Britain

Published by the Radio Society of Great Britain, 3 Abbey Court, Fraser Road, Priory Business Park, Bedford MK44 3WH

First published 2004

Third edition published 2013

© Radio Society of Great Britain 2013. All rights reserved. No part of this publication may be reproduced, stored in a retrieval system, or transmitted in any form, or by any means, electronic, mechanical, photocopying, recording or otherwise, without the prior written agreement of the Radio Society of Great Britain.

ISBN 9781 9050 8693 1

Publisher's note
The opinions expressed in this book are those of the author(s) and not necessarily those of the RSGB. While the information presented is believed to be correct, the author(s), publisher and their agents cannot accept responsibility for consequences arising from any inaccuracies or omissions.

Cover design: Kevin Williams, M6CYB
Production: Mark Allgar, M1MPA

Typography and design: Emdee Publishing

Printed in Great Britain by Charlesworth Press Ltd of Wakefield

Contents

1 Getting started **1**
What can be achieved? Do I need a big garden? Simple antennas. Receive antennas. Using the shack receiver. What you can hear on 136kHz. What you can hear on 472kHz. Obtaining a 472kHz transmit permit. Join the club.

2 Receivers **9**
What is needed. Suitable commercial receivers. Improving your receiver. Preselector, preamp and converter. 'SoftRock Lite' modifications.

3 Antennas and matching **23**
A tiny bit of theory. Antenna options. Marconi design considerations. Danger, high voltage! Using an existing antenna. Ground systems. Non-earth losses. Purpose built antennas. Marconi loading and matching. Transmitting loops. Practical loop antennas. Earth antennas. Why use 50 ohms on LF? Antenna supports.

4 Receive antennas **51**
Why use a receive antenna? Using a transmitting antenna. Positioning a receive antenna. Noise on feeders. Practical loop antennas. Active whip antennas. Reducing electrical noise.

5 Generating a 136kHz signal **67**
Signal-frequency VFOs. Crystals and ceramic resonators. Dividing a variable oscillator. HF transmitter as a signal source. Direct digital synthesis. GPS locking. Ultimate2 QRSS kit.

6 Transmitters **77**
Ready built and kit options. Second-hand equipment. Modifying audio amplifiers. Class D transmitters. Class E transmitters. G3YXM 136kHz 1kW transmitter. G0MRF 136kHz 300W transmitter. Low pass filter for 136kHz. GW3UEP's simple MF transmitters. G4JNT high power MF amplifier. Low pass filter for 472kHz. G4JNT medium power linear. Outline of an EER transverter. Power supplies.

7 Measurement and calculation **115**
Test gear for setting up the station. Dummy loads. Measuring a working station. Estimating radiated power. Field strength measurement. Scopematch tuning aid. 472kHz antenna tuning meter. Frequency stability and calibration.

continued >>

8   LF Propagation **129**
Ground waves. Sky waves. The effect of the Sun. Can we predict good conditions? Fading. Longer term changes. 472kHz.

9   Operating practice **141**
Operating on 472kHz. Operating on 136kHz. QRM and QRN. Remote receiving, grabbing and reporting. DX working. Operating away from home.

10   Modes used on 136 and 472kHz **155**
On-off keying modes. Multi-frequency modes. Modes requiring a linear transmitter. Better synchronisation. Modes compared. Interfacing with a computer.

10   VLF - below 30kHz **169**
What can be heard below 30kHz. Receiving equipment. Modes. Receive antennas. Permission to transmit. Transmitters. Transmit antennas.

- **Appendix 1: Information sources** **173**
Web sites. RSGB LF group. Publications.

- **Appendix 2: Components and software** **177**
Capacitors. Litz wire - is it useful? Ferrite and iron-dust cores. Component sources. Software.

Introduction

There are now more radio amateurs using the frequencies below 1MHz than ever before. More countries have become active on the 136kHz band (LF), but the main change has been the introduction at the ITU World Radio Conference 2012 of an international allocation between 472 and 479kHz (MF). This new MF band has been adopted by many countries, including the UK, but at the time of writing there are exceptions, including the USA and Russia. It is certain that activity on 472kHz will continue to grow in the coming years. Additionally, several European amateurs have obtained permission to carry out experiments around 10kHz (VLF), and have had some remarkable results.

This new edition of *LF Today* has been extensively revised, updated and expanded. There is significant material on the 472kHz band, including some very simple transmitters. The information on the older 136kHz band has been updated to reflect more recent operating practice. A new chapter describes and compares the various modes commonly used on low frequencies, including several created specifically for this part of the spectrum. For the first time a chapter shows how to receive and transmit on VLF. Alan Melia's chapter on propagation has been revised to show his most recent theories on how to predict when it is worth staying up all night to work DX.

Since it is necessary to build at least part of a low frequency station, *LF Today* has always had many construction projects, and this continues. However there are now more references to ready built equipment and kits. As these bands become more popular, it has never been easier to get on the air

The use of these bands varies from person to person, and there is something for everyone. There are DXers specialising in inter-continental contacts, listeners content to report the signals of others, QRPers amazed at their WSPR signals being heard hundreds of kilometres away whilst running impossibly low power, techies devising new data modes, constructors building from scratch, antenna experimenters practicing their alchemy to get a big signal from a tiny aerial, propagation researchers and ragchewers. What they share is a delight in using these unique bands, and achieving something that relatively few others have done.

As always, the aim of this book is to help the beginner to get started with the minimum of effort, and to help the experienced operator to get more out of his low frequency station. The material is a distillation of the wisdom and experience of the many experimenters who have reported on the RSGB LF Group over the best part of 20 years, combined with my own thoughts on the subject. If, as a result of reading *LF Today*, you get as much pleasure from the low frequencies as I have done, I will have succeeded.

Mike Dennison, G3XDV

Acknowledgements: Thanks go to all of those whose projects, words and pictures are incorporated into this book, in particular to Jim Moritz, M0BMU, who collaborated with me on the last edition and whose wisdom still permeates the pages; Andy Talbot, G4JNT, for his innovative projects and advice; Roger Plimmer, GW3UEP, for his simple 472kHz transmitters, Alan Melia, G3NYK for contributing the chapter on propagation, and Dave Pick, G3YXM, for checking the final draft. Lastly, several readers sent me corrections and comments on the last edition, all of which (I hope) have been addressed. If you have any comments on this edition, or spot any errors, please contact me.

introduction

1

Getting started

In this chapter:

- What can be achieved?
- Do I need a big garden?
- Simple antennas
- Receive antennas
- Using the shack receiver
- What you can hear on 136kHz
- What you can hear on 472kHz
- Obtaining a 472kHz transmit permit
- Join the club

THERE ARE CURRENTLY two international amateur allocations below 1MHz, an LF band at 136kHz and an MF one at 472kHz. Whether either band is permitted, and at what power levels, varies from country to country.

In the UK, the 136kHz band became available about fifteen years ago after a few years of experimentation on the unique, and now defunct, 73kHz band. The band extends from 135.7kHz to 137.8kHz, about the bandwidth of a single SSB station. Nevertheless, by using modern narrowband modes, 136kHz can accommodate many stations simultaneously.

In February 2007, UK licensing authority Ofcom agreed to issue some amateurs with licence Notices of Variation to experiment on frequencies between 501 and 504kHz. This part of the spectrum had been used for high-powered

Roger Lapthorn, G3XBM, in his QRP VLF/LF/MF station, which includes a 472kHz transverter (featured in this book) and the VLF beacon used with his earth antenna. Despite running just milliwatts or microwatts ERP, Roger says: "the lower bands have been a rich harvest of fun for me"

Morse code by ocean-going shipping for the best part of a hundred years, but by the start of the 21st century this activity had been rendered largely obsolete. Following internationally coordinated negotiations, this UK-only band was replaced by an internationally-agreed allocation from 472 to 479kHz.

Both the 136kHz and 472kHz bands have unique characteristics and are different from all of the higher frequency amateur bands. Propagation on these bands is very different to the HF bands (for more on this, see the chapter on propagation).

With worldwide contacts easily available on the HF amateur bands, some may wonder what is the attraction of operating at these low frequencies where it takes some effort to produce a station capable of a range of a few hundred kilometres.

As any microwave enthusiast will tell you, it is this effort and the overcoming of obstacles that makes 'difficult' amateur radio so much fun, and ultimately so satisfying. It is no coincidence that many LF experimenters have been established microwavers. People frequently report their first low frequency contact as the most exciting experience in years of radio operating.

A word of caution. Despite the enthusiasm engendered by operation in this part of the spectrum, it is possible to get discouraged. As there are few amateurs active compared, say, to the 20 or 40m bands, an efficient and versatile station is needed to stand a good chance of regular contacts. It is possible to make your first contact with a few watts of CW and a 'lash-up' antenna, especially on the 472kHz band, but many find it useful to develop their station with more RF power, a more efficient antenna and additional modes.

In the UK, and several other countries, the maximum power permitted is 1W ERP on 136kHz and 5W EIRP on 472kHz (see the chapter on Measurement for the difference between ERP and EIRP). To those more familiar with HF power levels, these appear to be very low limits, but being based on actual radiated power the antenna efficiency is taken into account. With typical practical antenna efficiencies of in the region of 0.01% on 136kHz and 0.1% on 472kHz, RF inputs of around 1kW and 500W respectively are needed to achieve the maximum licensed power. As on the HF bands, most operators run much less than the power limit. Home-made transmitters are frequently used, but a few commercial rigs are now available.

Since it is much simpler to radiate an effective signal at 472kHz than at 136kHz, many stations will try that band first. This book will show you how easy it is to do that, but it will also demonstrate how to go the whole distance and experience our lowest amateur band as well. The extra work involved in doing so will pay dividends, not only on 136kHz itself, but in developing techniques that will also improve your results on 472kHz.

What can be achieved?

At the time of writing, the 472kHz band is quite new and not every country has permitted its use. Nevertheless, Europe-wide contacts have been made routinely during darkness hours and special experimental stations in the USA have been received in western Europe. Using the WSPR2 mode an Australian station has been logged in France. Daytime ranges are a few hundred kilometres.

The typical range of a suburban station on the 136kHz band depends not only on the radiated power but also the type of transmission used. Conventional CW, typically at 12WPM, can provide contacts up to a 1000km or so. Data transmissions using CW bandwidths may work over a similar range. Distances in excess of 1000km range usually require specialist modes such as QRSS, Opera and WSPR (see a later chapter for details) and these modes comprise the majority of activity. Daytime (ground wave) ranges are greater than on 472kHz. Long distance contacts by sky-wave are available at night.

As with any amateur radio, the number of contacts and the distances involved will depend on the amount of time and effort expended, propagation conditions and transmitted power levels.

Do I need a big garden?

An efficient half-wave dipole for the 472kHz band would be over 300 metres long and well over 50m high. Even a quarter-wave ground plane would require a 150m tower and an earth system nearly half a kilometre in diameter. On the 136kHz band, the situation is far worse!

Needless to say, some compromises are needed to operate on these bands in a housing estate. Fortunately, it is possible to achieve usable efficiencies with 'domestic' antennas, especially on 472kHz, provided care is taken. It is helpful to have enough real estate to accommodate, say, an 80m dipole some 10m high, but success (even at 136kHz) has been achieved from very small gardens, and even the roof of an apartment block. Several of the early experimenters on 500kHz successfully used HF dipoles strapped as Marconi antennas and tuned against ground.

Simple antennas

Two basic types of antenna have been used for amateur LF/MF transmitting: the loop and the Marconi. These will both be dealt with in detail later, but for the purposes of this chapter let's see how easy it is to erect a Marconi antenna.

Guglielmo Marconi discovered that a single vertical element would work as an antenna when a connection was made to ground to complete the circuit (**Fig 1.1**). He also found that by folding the top of the vertical parallel with the ground, the height - and hence the cost of the mast - could be reduced with very little reduction in efficiency. The resultant inverted-L and T antennas (**Fig 1.2**) are familiar to those who operate on the 160 and 80m amateur bands.

For our case, even an inverted-L is too big. Fortunately, a Marconi antenna will work when its physical length is much less than a quarter wavelength, provided it is properly matched. Naturally, this will reduce its efficiency.

An established amateur radio station may well already have the basis of an LF Marconi antenna. A dipole for 160, 80 or even 40m can be adapted by connecting the vertical part of the coax as a single

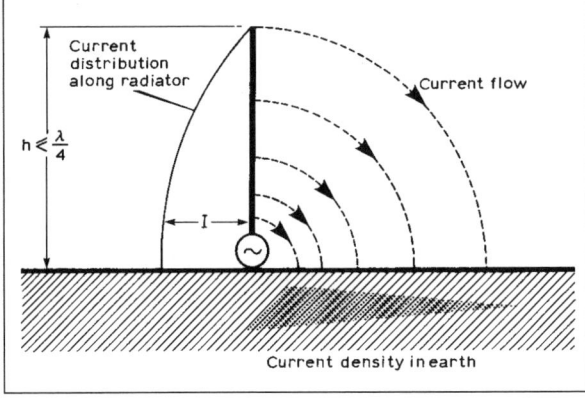

Fig 1.1: Current distribution of a short radiator over a plane earth

LF TODAY

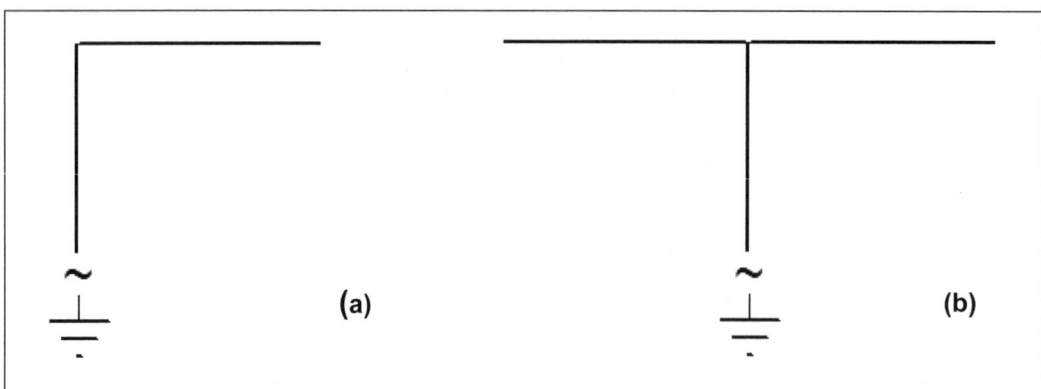

(above) Fig 1.2: Reducing the height needed for a low frequency vertical. (a) An inverted-L and (b) a 'T'

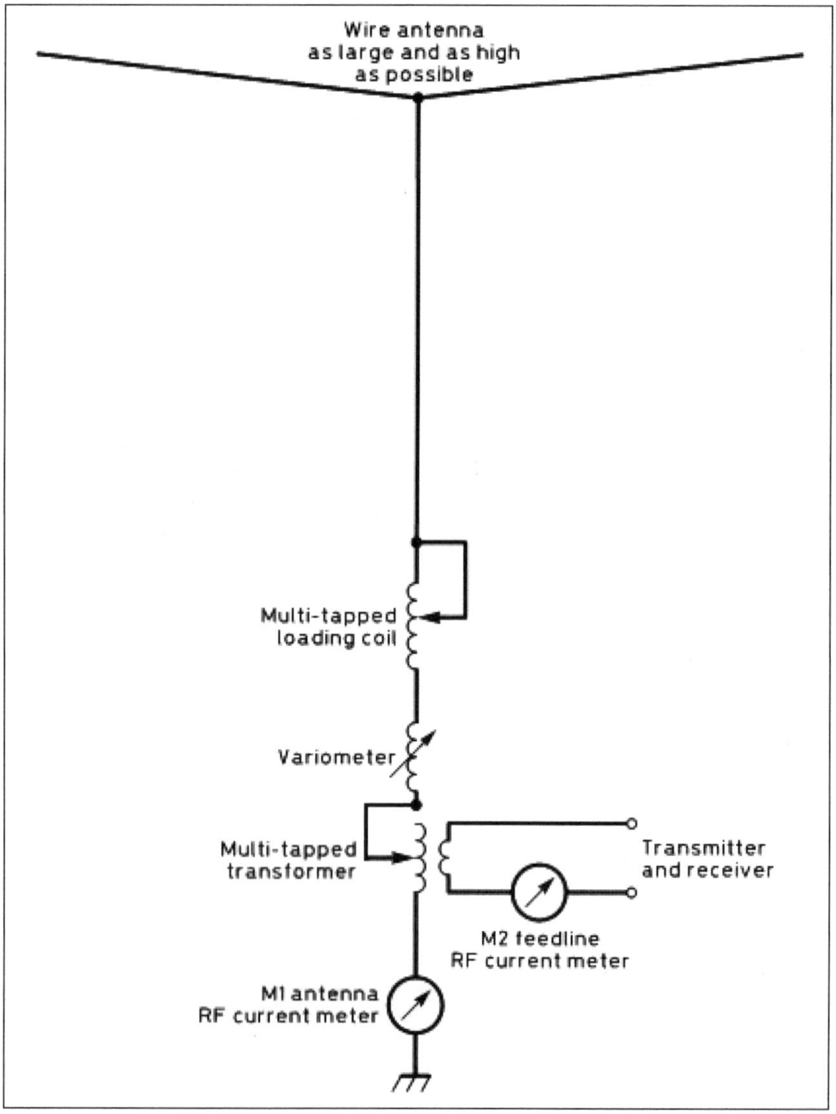

Fig 1.3: The typical arrangement of an LF antenna. The main element could be an existing dipole with the feeders connected together, or it could be purpose-designed. By adjusting taps on the loading coil and matching transformer, the antenna can be used on both 136 and 472kHz

wire and paying additional attention to insulation (see the antennas chapter). Even a G5RV could be used in this way. A tower or pole mounted HF or VHF antenna system might also be pressed into service.

If started from scratch, an LF/MF Marconi should be designed to have as long a vertical section as possible, and a top section giving as much capacitance to ground as possible without compromising on height (**Fig 1.3**). This need not be in the classic L or T shape. Anything that produces capacitance at a good height will work.

Since the resultant Marconi will be very short compared with the required electrical wavelength, the feedpoint will have considerable capacitive reactance. This must be compensated for (tuned out) by inserting an inductance - perhaps several millihenries - in series with the antenna lead.

A good ground connection is also required. For initial low power experiments, a single earth stake or a water pipe is a good starting point.

Although the theoretical feed resistance of a very short Marconi is a fraction of an ohm, this is greatly increased by the resistance of the coil together with earth and other losses. Your LF antenna may have a feed impedance much closer to 50 ohms than you would think.

Receive antennas

Even before considering transmitting, much can be learned by listening on the bands. Unfortunately, it can also a really good way of putting yourself off the low bands for life. This is because an unmatched wire plugged into a typical HF transceiver will often produce a lot of noise and no signals. No matter what antenna is used for your first excursion onto low frequencies, it is important that it is resonated and (if it is a Marconi type) tuned against ground.

In the main, resonating involves winding a coil that can be adjusted from about one to three millihenries (perhaps 200 to 400uH for 500kHz). A proper radio earth is required - do not rely on the receiver being connected to earth through the power unit's mains plug.

An alternative, and often effective, receive antenna can be made by winding several turns of wire round a frame, parallel tuning this with a variable capacitor from an AM broadcast radio and coupling it to the receiver with a single turn loop. The antenna need not be placed high up, but must be mounted away from household wiring.

A multi-turn receiving loop antenna for 136kHz, built by F6AGR

LF TODAY

If you have a TS-850, you are already well on the way to hearing amateurs on the low frequencies

Using the shack receiver

Many modern HF transceivers tune down to around 100kHz, which theoretically makes them suitable for reception at 136kHz. However, the majority have greatly reduced performance at LF, often leading to early disappointment for the casual listener. Some radios attenuate MF signals in order to reduce interference effects from Medium Wave broadcast stations, and this can affect reception at 472kHz.

A useful HF transceiver for 136kHz reception is the Kenwood TS-850 which performs very well and is available these days at a reasonable price second hand. Its Medium Wave attenuator switches in above 500kHz, so the radio receiver works well on the 472kHz band.

The sensitivity of many radios can be greatly improved by using a pre-amplifier (see the receivers chapter), but this must incorporate tuned circuits to avoid amplifying huge non-amateur signals which may well result in blocking or mixing effects.

Dedicated general coverage receivers, for example the AOR7030, are capable of good reception at both low frequency bands. Some modern software defined radios (SDR) can also perform well at these frequencies.

Suggestions for suitable radios, as well as pre-amplifier circuits, can be found in the receivers chapter.

For reception of narrowband modes, such as QRSS30 or WSPR15, the receiver's oscillators must have minimal drift. Because of the very strong utility stations close to the 136kHz band, and because both bands are very narrow, it is highly desirable to fit a CW filter in the receiver's IF. A receiver with very good dynamic range will also help.

What you can hear on 136kHz

Like many amateur bands, 136kHz is shared with utility stations and is also adjacent to them. **Fig 1.4** shows the European 136kHz environment.

It can be seen that DCF39, just above the band on 138.830kHz, is a potent signal which can be the source of receiver problems, but also makes a most useful beacon. This 50kW ERP transmission from Magdeburg, Germany, comprises a single carrier, interrupted every ten seconds or so by a data burst lasting about one second. A similar sounding signal, from Budapest, can be found on 135.43kHz; sidebands from its data bursts can spill into the lower part of the amateur band. Not only are these an excellent starting point for finding the band and testing receiver and antenna improvements, but they also make good propagation indicators for stations outside Europe looking for DX signals. Beware, though, that these stations are very much stronger than any amateur signal you are likely to hear.

Within the band, around 135.8kHz, is SXV. Originating in Greece, SXV transmits teletype continuously. In the UK, it is audible during daylight hours on a good receive set-up, but gets very strong at night when the signal is propagated by skywave.

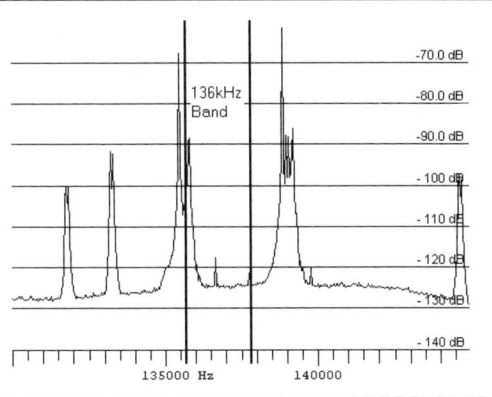

Fig 1.4: The 136kHz amateur band as received in the UK, showing the close-in very strong utility stations

Amateur CW signals are relatively rare, but can be heard in the middle of the band, from just above 136.0kHz to just above 137.0kHz. QRSS signals, which can be identified by ear as several seconds of carrier followed by several seconds of silence, are usually around 137.7kHz and occasionally close to 136.3kHz during inter-continental tests. Computer-based modes such as Opera and WSPR, and experimental transmissions. may be found between 137.4 and 137.7kHz (see operating chapter for more details).

The best time to listen for amateur signals is at weekends. With a good set-up, you should be able to hear CW signals from the UK and Western Europe, and read QRSS, Opera or WSPR signals from as far as Russia and even North America. More about DXing later.

What you can hear on 472kHz

The first thing many people notice on this band are aeronautical non-directional beacons (NDBs) in central Europe, which can be heard in darkness hours transmitting their two- or three-letter callsigns in Morse every few seconds. They run around 100 watts to short loaded verticals and transmit MCW which, like AM, produces a central carrier and weaker sidebands about 1kHz either side of it containing the Morse. A list of the NDBs most frequently heard in the UK is shown in the chapter on Operating. Strong local NDBs adjacent to the band can be used during the initial testing of equipment or antennas.

Amateur stations are active mostly at weekends and in the evenings. CW is popular in the lower half of the band, with Opera, WSPR and other modes commonly used in the top half. At the time of writing there is no formal bandplan for the 472kHz band; check with the RSGB LF Group [1] to find the centres of activity for each mode.

Although this band is an international allocation, each country must add it to their licence and it will take some time before this is done in every country.

Obtaining a permit to transmit on 472kHz

At the time of writing, the UK licence does not automatically include 472kHz. However, it is easy for Full Licensees to obtain permission to use the band. Just fill in the form on the RSGB web site [2] and within a few days a Notice of Variation (NoV) to your licence will arrive. Unlike the old 500kHz permit there is no need to provide information about your station. The permitted power level in the UK is 7dBW (5 watts) EIRP. Other countries may have a different (often lower) limit. We deal with measuring or calculating your EIRP in a later chapter.

An important requirement of the NoV is not to "cause interference to stations operating in the aeronautical radionavigation service or on 490kHz to the maritime mobile service". Furthermore you cannot claim protection from interference to your station "from other wireless telegraphy or electronic equipment".

Join the club

If this chapter has got you interested, the next step is to join the RSGB LF Group which is a forum on Yahoo. After signing up (free), you will receive in your e-mail inbox a copy of every e-mail sent to the group, and you will be able to post messages to the group. Alternatively you can opt to receive a daily digest of emails, or to read and post on the Yahoo web site. There is news and comment, plus technical discussion and advice. This is where you can ask questions on LF matters, no matter how simple or complex, and you will get replies from experienced, knowledgeable and friendly people. To join the Yahoo group, simply send a blank e-mail to: *rsgb_lf_group-subscribe@yahoogroups.co.uk*, or visit the web site at *http://uk.groups.yahoo.com/group/rsgb_lf_group/* and follow the appropriate link. You should then receive several LF-related e-mails each day.

References

[1] *http://uk.groups.yahoo.com/group/rsgb_lf_group/*
[2] *http://www.rsgb.org/operating/novapp/nov-472-479-khz.php*

2

Receivers

In this chapter:

- What is needed
- Suitable commercial receivers
- Improving your receiver
- Preselector, pre-amp and converter
- 'SoftRock Lite' modifications

PROVIDED YOU HAVE A sufficiently sensitive and selective receiver (see below) it is easy to find amateur signals on the 472kHz band. These may be CW or data but phone modes are not encouraged on such a narrow band.

On the 136kHz band the amateur signals can be extremely weak, whilst very strong non-amateur transmissions are present on the antenna. This leads to the conflicting requirements of good sensitivity and resistance to blocking.

Frequency stability is an issue when receiving very slow QRSS, and a good noise blanker can help during the summer months when static dominates.

What is needed

Tuning the antenna to resonance is vital for transmitting but easy to overlook when just receiving. Amateur stations may be received at 472kHz on an untuned antenna, but better results will be obtained when the system is properly resonated. On 136kHz, however, tuning the antenna is essential to be able to receive any but the strongest amateur stations. Some antenna tuning arrangements are shown in **Fig 2.1**.

Antennas for transmit/receive and just for receiving are detailed in later chapters.

It is important to site the antenna away from sources of local interference. This topic is discussed in detail in the chapter on Operating Practice, but for the moment you need to be aware that local electrical noise can be a major problem on both bands.

Many amateur HF transceivers will receive well on the 472kHz

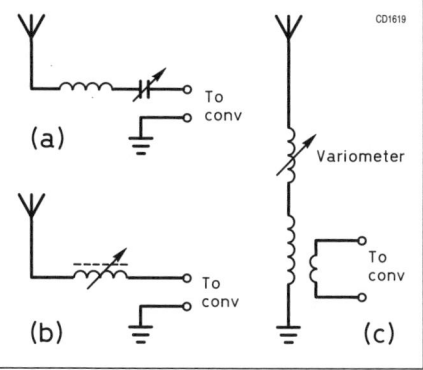

Fig 2.1: Methods of tuning a typical LF or MF antenna: (a) and (b) are adequate for receiving, whilst (c) is typical of a transmitting installation. An earth connection must be used with these systems

band, though strong Medium Wave broadcast signals may be an issue if an untuned antenna is used.

On the 136kHz band a radio with poor sensitivity is capable of receiving non-amateur transmissions, and this may lead the beginner to believe it is satisfactory. Unfortunately, amateur signals are several tens of decibels closer to the noise level. For instance, DCF39 on 138.830kHz should be well over S9 on your s-meter if amateurs are to be received satisfactorily.

A good rule of thumb is to have a system capable of hearing a constant background crackle in the absence of signals (and thunder storms) - this is the so-called 'presence' that will be familiar to experienced HF operators. There should be a significant reduction in the noise level when the antenna is disconnected.

Ideally, your receiver will be capable of sufficient sensitivity without modification (see below) but in many cases sensitivity can be improved by the addition of a tuned or filtered pre-amplifier (see below).

Plainly, good dynamic range is important in a 136kHz receiver, as is proper gain distribution. An RF gain control or input attenuator is desirable. Requirements for 472kHz are less stringent provided you do not have local amateur signals and provided you have sufficient front end (or antenna) selectivity to reduce the amplitude of strong medium wave broadcast stations. On both bands it is useful to be able to switch the AGC off if required.

The modes used are almost all narrowband. For CW and QRSS the receiver should have a narrow IF filter, or at least provision for fitting one. Computer-based modes, such as WSPR and Opera, need an SSB-width filter and use their own built-in DSP to provide selectivity. On the 136kHz band, a steep-sided SSB roofing filter with a good stop-band is desirable to reduce the impact of the adjacent non-amateur stations which can be 60dB or more above the level of readable amateur signals, with a frequency separation of less than 1kHz.

For CW or QRSS3 operation, most radios will have adequate frequency stability. If your interests lie in digital modes or in working DX using extremely slow CW (eg QRSS30), a high level of stability is needed, such as that.achieved by fully synthesised receivers

For some specialised narrow band operating modes, a higher order of stability is required. This has been achieved by using high stability synthesiser reference frequency sources, such as TCXOs, OCXOs, and even atomic clock or GPS-derived frequency standards.

Frequency readout is often confused with frequency stability, but a readout to 1Hz does not mean the radio will stay within 1Hz for long periods. Conversely, some radios will have excellent stability but have a readout in 10Hz steps. Again, this is of no consequence for CW operation, but for QRSS the ability to set the receiver accurately to 1Hz can be useful. Note, though, that provided the receiver is stable, it is possible to use software to improve on a poor frequency display - see the chapter on Measurement and Calibration.

Lightning static is audible for most of the time at LF as there is usually a storm of some kind within 2000km or so. This dramatically increases in the summer months, leading some to abandon the low bands during this period. A really good noise blanker can be an asset under these conditions, though few commercial radios have this.

CHAPTER 2: RECEIVERS

Suitable commercial receivers

Most modern transceivers cover the 136 and 472kHz bands, although not all have sufficient sensitivity at low frequencies. Some work better on 136kHz than on 472kHz where an attenuator may be fitted to reduce blocking on the Medium Wave broadcast band, whilst others will perform better at the higher frequency. Most will work fine with the addition of a pre-amplifier for one of both bands (see later). Dedicated receivers, including software designed receivers (SDRs) often perform well.

When using a receiver on 472kHz, it may be useful to reduce the level of strong Medium Wave transmissions by using a low-pass filter such as the one shown in the Transmitters chapter.

Amateur HF transceivers

Rather than use your main HF station receiver, it can be useful to buy a cheap second-hand receiver or transceiver solely for use with your LF/MF station. It is important, though, to choose the right one. In any case, since amateur transceivers are not designed for the bands below 1.8MHz, there is little relation between the HF performance, cost, or sophistication of a particular model, and the sensitivity at lower frequencies. Therefore it may well be that older, cheaper models perform better at LF than their newer successors.

A good performer on 136kHz is the Kenwood TS-850, but don't assume that all of the more recent Kenwood models are an improvement on this band. The '850 has excellent sensitivity and a clean DDS. A range of CW filters can be fitted both at 8MHz and at 455kHz, allowing very high skirt selectivity. It also performs well at 472kHz. A disadvantage is that the minimum tuning step is 10Hz, which makes very slow QRSS modes more complicated to set up.

The Kenwood TS-440 is also very sensitive and, provided the internal 20dB attenuator is used, the intermodulation behaviour is very good. The noise blanker in the TS-870 is reported to be effective, though its LF sensitivity is not as good as the TS-850. The receiver is ideal for use with large antennas as is the TS-140.

Some transceivers, such as the TS-930 have a transverter port making it easy to get a fully-featured LF/MF radio in conjunction with an external home-made transverter (see the chapter on Generating a Signal).

The Icom IC-706 (Mk1 and Mk2) has excellent stability and the frequency can be set in 1Hz steps. However, it is insensitive at 136kHz and benefits greatly from an LF pre-amplifier. Used on 472kHz with a large resonant antenna, sensitivity is adequate with no sign of broadcast band interference. The AGC

> Note: The advice on commercial receivers given in this chapter is based on user reports and is intended purely as guidance. Actual performance may vary from one equipment to another. Potential buyers are advised to make their own tests on the radio before parting with their money.

The cheaply available IC-706 is stable with frequency readout down to 1Hz. The receiver is good for 472kHz but needs a pre-amplifier to be efficient on 136kHz

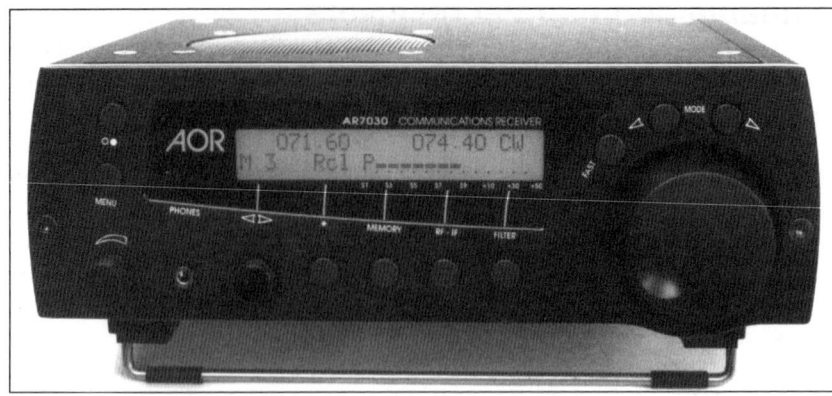

The AOR7030 is a receiver that performs very well at low frequencies

cannot be switched out but there is an RF gain control. A fixed audio level is available on a rear socket which is handy for connecting to a computer sound card for QRSS and data modes. The noise blanker is totally ineffective at LF. Similar performance has been experienced with the IC-718 and the IC-756 PRO. The IC-751A has been used successfully on 136kHz after the addition of an AF filter, although the frequency stability has been reported as inadequate for QRSS. Other LF operators use the IC-761, IC-765 and IC-781, adding CW filters in both IFs.

From Yaesu, the FT-990 is very sensitive and has built-in DSP to supplement the very good optional CW filters; the AGC can be switched out. The FT-1000MP requires an external preamplifier for LF, as does the FT-817 which may also not be stable enough for very slow QRSS speeds.

Dedicated receivers

Stand-alone receivers are more likely to perform well at low frequencies. The Yaesu FRG100 is reported to be sensitive and stable, and as good as the TS-850. Tuning well below the 136kHz band, the AOR7030 is useful for LF work with good sensitivity, 10Hz tuning steps and optional 300Hz or 500Hz IF filters. JRC's NRD-345 is sensitive and stable, and has several optional narrow filters; the NRD-525, 535 and 545 are described as very good and the NRD-91 tunes down to 10kHz. The Icom IC-R75 can be fitted with an oven controlled oscillator and is very sensitive; the AGC can be turned off. Lowe's HF-150 has good IMD performance, but no CW filter, RF gain or AGC off switch; the HF-225 has a 200Hz filter at AF. Other receivers that have been used at LF include the Cubic R3030 and R3090.

Receivers intended for commercial operation, such as those made by Racal, Plessey, Harris, Collins and others, are also likely to perform well at LF/MF, though some may have keyboard entry rather than a dial for tuning. Most have high stability oscillators tunable in 1Hz steps, inputs for a high stability frequency reference and AGC that can be switched off. They will also have outstanding IMD and cross-modulation performance. The Racal 1792, which has IF passband tuning, has been popular with UK amateurs using the LF/MF bands. The older RA1772 also performs well. Another useful receiver is the Harris RF590 which has good sensitivity and an effective noise blanker. From Germany, the EKD300 is reportedly a good performer.

SDR receivers

Software defined radio (SDR) is now becoming part of the mainstream of amateur radio, with both home constructed and commercially produced SDR hardware and software now widely available. PC-based spectrogram software has been used for several years in conjunction with conventional receivers for the 'visual' LF/MF operating modes such as QRSS and narrow-bandwidth data modes; SDR is the natural extension of this trend.

Homebrew amateur SDR projects most commonly use PC-based digital signal processing software, using the PC sound card for A/D conversion of the incoming signal. Since the sound card is usually limited to 48kHz sample rate, the maximum signal frequency that can be handled by the sound card input is 24kHz. So for amateur band use, some form of external down conversion is required. This generally takes the form of an I/Q down converter, with in-phase and quadrature outputs feeding the left and right stereo inputs of the sound card. The I/Q signal format permits image rejection to be performed by the SDR software, and also extends the bandwidth that can be processed by the sound card to 48kHz; this is ample to cover the narrow amateur 136kHz and 472kHz bands with fixed, crystal-controlled conversion frequencies. All required tuning, filtering and demodulation functions are then performed in the digital domain by the SDR software [1, 2, 3]. This results in a very simple yet capable amateur band receiver; modifications to the well-known KB9YIG 'SoftRock' SDR receiver kits to permit 136kHz and 472kHz reception are described later in this chapter.

General coverage, direct-digitising SDR receivers are now also commercially available to amateurs at reasonable prices, and these commercial SDRs are now the receivers favoured by dedicated low frequency operators. These include the RFSpace Inc SDR-IQ [4], the Perseus SDR receiver [5], and the Afedri SDR-Net [6]. These receivers are supplied with their own native SDR software, but can also be used in conjunction with other software in order to generate high resolution spectrograms for 'visual modes' operation.

Less expensive SDR solutions are available, such as the Funcube and even the DVB-T dongles that can be obtained really cheaply from internet suppliers.

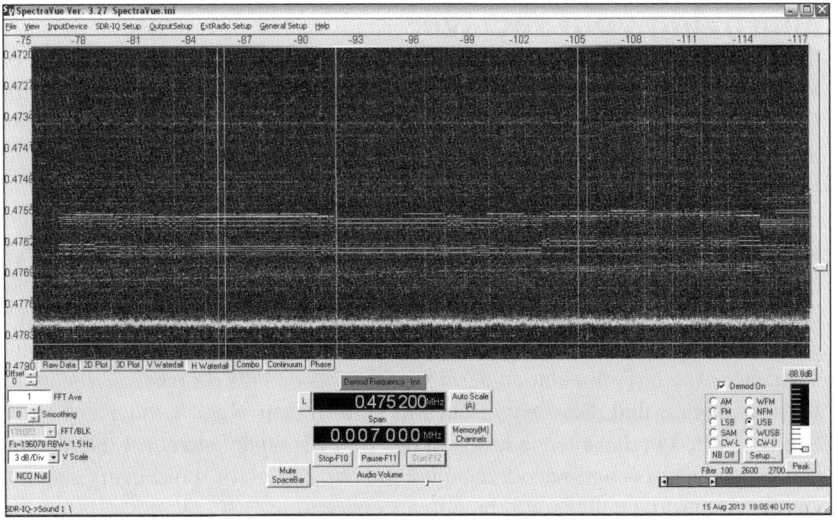

The entire 472kHz band displayed on an SDR-IQ using the Spectravue software. The wobbly line at the bottom is local interference from an LCD television

However, the same care needs to be exercised in choosing an SDR as is needed with a 'conventional' receiver, and attention must be paid to frequency stability, sensitivity at the required frequency, and dynamic range.

Vintage receivers
Some 'boat anchor' radios tune to 472kHz and even as low as 136kHz. These do not have the 'brick-wall' selectivity required to avoid the very strong utility signals adjacent to the 136kHz band, but may well have adequate sensitivity for your first attempts at receiving on 472kHz. Frequency stability is likely to be a problem for modes other than CW.

Selective measuring sets
Used for making measurements on the old analogue telecommunications systems, selective measuring sets (SLMS, or sometimes TMS) can make good LF receivers. They almost all cover frequencies down to VLF and some include the HF spectrum, too.

The advantages of going for this type of receiver are: adequate sensitivity for amateur use, good dynamic range, a calibrated input attenuator, an output meter with a readout to 0.1dB, probably a very narrow IF filter (less than 100Hz), good stability, constant sensitivity over a wide range of frequencies and the possibility of a matching oscillator that can be locked to it. They are also ideal for making accurate field strength measurements (see the Measurements chapter).

On the debit side are: no AGC, sometimes no SSB or CW demodulator (this can often be overcome by connecting your station receiver to an IF output socket), often no intermediate-bandwidth filter (just speech bandwidth and sub-100Hz) and a greater size and weight than most amateur specification radios.

Since an SLMS can be picked up second hand (check eBay regularly) for £100 or less, it is often a good investment. Look for the East German MV61 or 62 'Pegelmesser', Siemens D2155, Wandel and Goltermann (W&G) SPM-3, SPM-12, SMP-30, SPM-60 or similar.

More on using these sets as LF receivers can be found at [7, 8].

Improving your receiver
Ideally, you should always start with an antenna that is tuned to resonance or at least tuned circuits ahead of the receiver (see the chapters on antennas for transmitting and receiving). The practical effect of resonating the antenna can be dramatic.

Additionally, many receivers will benefit from a preamplifier. However, do not under any circumstances use an untuned circuit. In addition to amateur signals, this will amplify non-amateur transmissions such as MSF on 60kHz, the BBC on 198kHz or medium wave broadcasters, which will lead to an increase in noise and intermodulation products. Too much gain should be avoided, so a gain control or switched attenuator before the amplifier would be useful.

To check the intermodulation performance of your 472kHz receiver, listen for spurious signals that disappear or become significantly weaker when an attenuator is inserted in the antenna lead. On the 136kHz band, listen on 138.000kHz. If you can hear a rough-sounding signal pulsing on and off once every second, you have a problem. This will be a mix between the BBC on 198kHz and the

Anthorn time signal on 60kHz. The fix is better gain distribution - probably less front-end gain - and better input filtering. Note that late at night it may be possible to hear noise modulated by broadcast signals close to the top or bottom of the 136kHz band. This is ionospheric mixing - the so-called 'Luxembourg effect' - and cannot be eliminated by improving your receiver.

Turning down the RF gain control may well produce a better signal to noise ratio. In some cases, especially for QRSS operation, it may be useful to turn off the AGC - note that some receivers do not allow this.

When using CW or QRSS, any serious operator will invest in a CW filter, preferably the best available and if possible fitted to both IFs. Since there is little Morse above 12WPM on these bands, at least for DX working, an IF filter of 250Hz can be fitted. An analogue or DSP audio frequency filter can be a helpful addition, but is not an adequate replacement for good IF filtering. .

Many datamodes now require the receiver to be set to an SSB bandwidth, and at first glance it seems foolhardy to operate on LF/MF using an IF bandwidth that covers a large chunk of the entire amateur band! However, provided there is good protection against strong out-of-band signals, this works just fine. These modes (eg Opera or WSPR) contain extremely narrow DSP filters within their software and they perform better when presented with audio of SSB bandwidth.

What if your trusty HF radio has all of the above features, but does not tune down to 136kHz, or performs badly at 472kHz? All is not lost, as a properly designed LF converter will extend the frequency range. More of that later.

M0BMU preselector, preamplifiers and converter

The circuit blocks described below have been in use at M0BMU for some time, and combined together in various configurations have proved to be a versatile interface between a wide range of receiving antennas and receivers to provide good receiving performance in the LF/MF range.

For many receivers of good to 'average' sensitivity at LF, the preselector will be sufficient by itself when used with reasonably large loop or wire antennas to bring the band noise above the receiver noise floor. At M0BMU, the preselector has been used like this with a Racal RA1792 receiver, and an Icom IC718 transceiver.

For receivers with very poor sensitivity, for example the Yaesu FT817 at 136kHz, adding the 20dB preamplifier after the preselector will give a sensitivity improvement, however the the combination of preselector-plus-converter gives better results, with the receiver used as a HF tunable IF.

Small tuned loop antennas benefit from the 20dB preamplifier, used directly with a reasonably sensitive receiver, or to feed the converter. If additional gain and filtering is required, this can be achieved by connecting the preselector between the preamplifier output and the input of the converter or receiver.

Preselector for 136kHz and 472kHz

The basic preselector circuit is shown in **Fig 2.2**. It provides an antenna tuning and filtering function, along with considerable gain to drive a low-impedance receiver input. The tunable L-C input circuit has a narrow bandwidth of a few kilohertz, which substantially reduces the level of unwanted signals at the receiver input, particularly medium wave broadcast stations.

LF TODAY

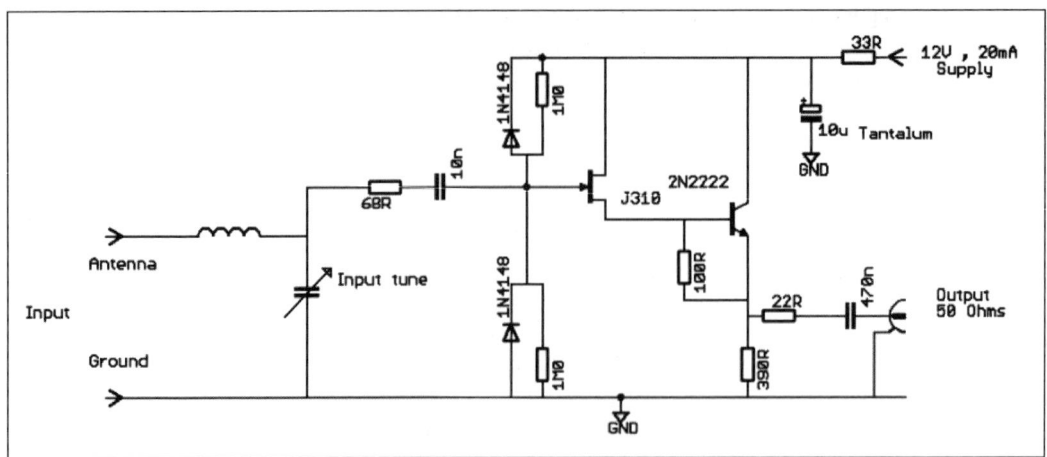

Fig 2.2: LF/MF Preselector - input connections for different antennas are shown in Fig 2.3. The antenna input components are 1mH and 500pF for 472kHz, and 3.3mH and 1000pF for 136kHz

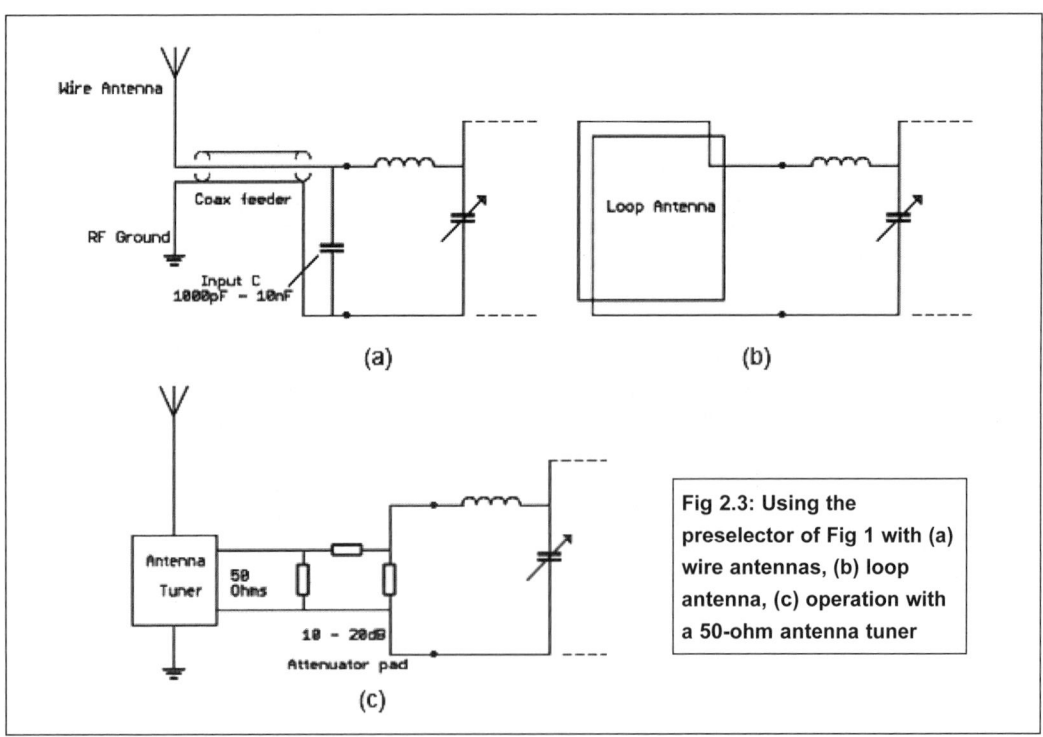

Fig 2.3: Using the preselector of Fig 1 with (a) wire antennas, (b) loop antenna, (c) operation with a 50-ohm antenna tuner

A compound FET/bipolar unity-gain buffer amplifier provides a low output impedance to drive the receiver, with a very high input impedance to minimise loading of the tuned input circuit. Although the follower has a gain around unity, the circuit has considerable overall voltage gain due to the impedance step-up of the input tuned circuit.

The input circuit can be adapted for use with wire antennas, or various types of loop antenna, as shown in **Fig 2.3**. For use with a wire antenna, a shunt capacitor is added to the input circuit (**Fig 2.3(a)**). The value of this capacitor controls the impedance transformation, and so acts as a sort of RF gain control - larger input capacitors result in lower signal level at the preselector output. This capacitor should be selected to provide the minimum gain that achieves adequate signal-to-noise ratio. For short wire antennas of 5 - 10m, a 1000pF capacitor is typically about right, whilst up to 10nF can be used with larger long wire antennas. A wire antenna may be connected to the preselector input through a coax feeder, in which case the capacitance of the coax cable makes up part or all of the input capacitor. This allows the antenna feed point and earth connection to be well away from the shack, which can help to reduce noise pick-up.

For use with un-tuned loop antennas (**Fig 2.3(b)**), the shunt input capacitor is omitted, and the loop connects directly to the input, effectively forming part of the tuning inductance. This arrangement gives quite a large voltage step-up, and the overall voltage gain can be 30dB or more, boosting the generally low output from the loop antenna. The antenna may be a large single turn of wire with an area of $10m^2$ or more, or a smaller, multi-turn loop.

A 1m x 1m, 10 turn square loop has been used satisfactorily with this circuit for 136kHz reception, with five turns used for 472kHz. Loop antennas may also be connected to the preselector input via a considerable length of coax, allowing convenient tuning from the operating position.

The preselector may also be used in conjunction with an existing transmitting-type antenna and tuner providing a 50 ohm match, in order to provide additional gain and filtering. In this case, the signal level at the antenna tuner output will probably be quite large, and an attenuator pad can be inserted between the antenna tuner and the preselector (**Fig 2.3(c)**) to reduce the signal level and also reduce tuning interaction between preselector and antenna tuner. The input as shown in Fig 2.3 can also be connected directly to a low impedance source, such as another preamplifier, or an active antenna. Used in this way, it functions as a tuned preamplifier with around 30dB gain.

The preselector circuit can be used over a wide range of frequencies by choosing appropriate values of input tuning inductor and capacitor. Values in Fig 2.3 are given for 136kHz and 472kHz, and a band-switched version of the input tuning circuit covering 10kHz to 600kHz in five overlapping bands is shown in **Fig 2.4**. The band switch selects a tuning inductance between 1mH

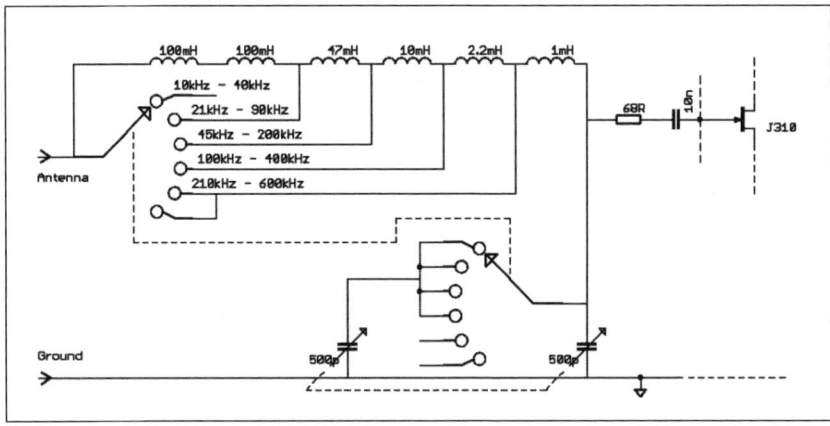

Fig 2.4 10kHz - 600kHz multi-band input circuit for preselector

LF TODAY

Fig 2.5: 20dB preamplifier

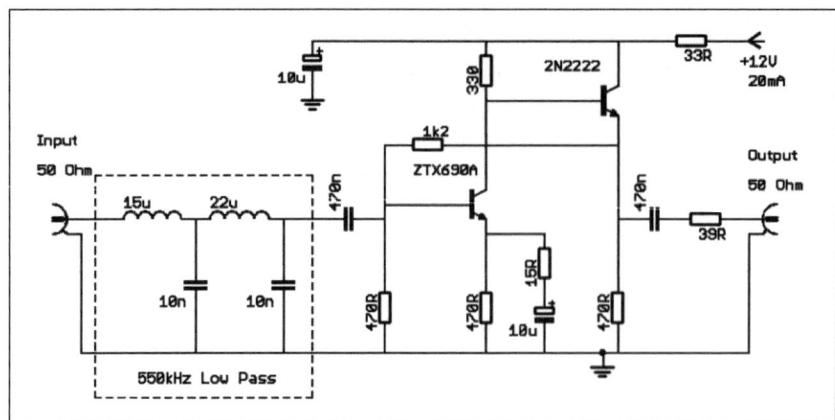

and about 250mH, which is resonated by both gangs of a 500pF + 500pF tuning capacitor, except on the highest band, where only one gang is used.

The inductors used in the prototype were small radial-leaded filter chokes wound on a ferrite bobbin (eg Panasonic ELC series, Wurth Electronics WE-TI series, available from RS components [9]). The inductance values are not critical, and other types of inductor with a Q of 50 or more at the receive frequency should be satisfactory.

Pre-amplifier

For particularly insensitive receivers, more gain may be required than is provided by the preselector by itself. Also, small tuned-loop antennas have quite low signal output and require a low-noise preamp with 50-ohm input and output impedance to drive the receiver. The preamplifier circuit in **Fig 2.5** gives about 20dB of gain through the VLF, LF and MF ranges. The ZTX690A transistor gives a noise figure of about 3dB in the LF/MF range in this circuit; if low noise is not so important, other transistors such as the 2N2222 can be used with a few dB increase in noise figure. The preamp is shown with an optional low-pass input filter with cut-off frequency around 550kHz; this is advisable in order to reduce the level of broadcast signals reaching the receiver.

LF/MF to HF converter

Fig 2.6 shows the converter. It uses a 4MHz crystal oscillator and broadband diode mixer module to up-convert input signals from a few kilohertz up to the 550kHz cut-off frequency of the input low-pass filter, to an output range of 4.00MHz to 4.55MHz.

The LF/MF input signal is fed into the DC-coupled IF port of the SBL-1 mixer, and the HF output is taken from the RF port - this allows input frequencies below the 500kHz minimum of the RF port.

The crystal oscillator uses one gate of a 74HCU04 hex CMOS inverter IC, with the remaining five gates used as a buffer amplifier to drive the diode mixer. A wide range of other crystal frequencies could also be used to obtain different output frequency ranges if preferred; due to the broad-band nature of the mixer, no further modification is needed other than to change the crystal. An oscillator frequency below 2MHz makes the circuit more susceptible to IF breakthrough

CHAPTER 2: RECEIVERS

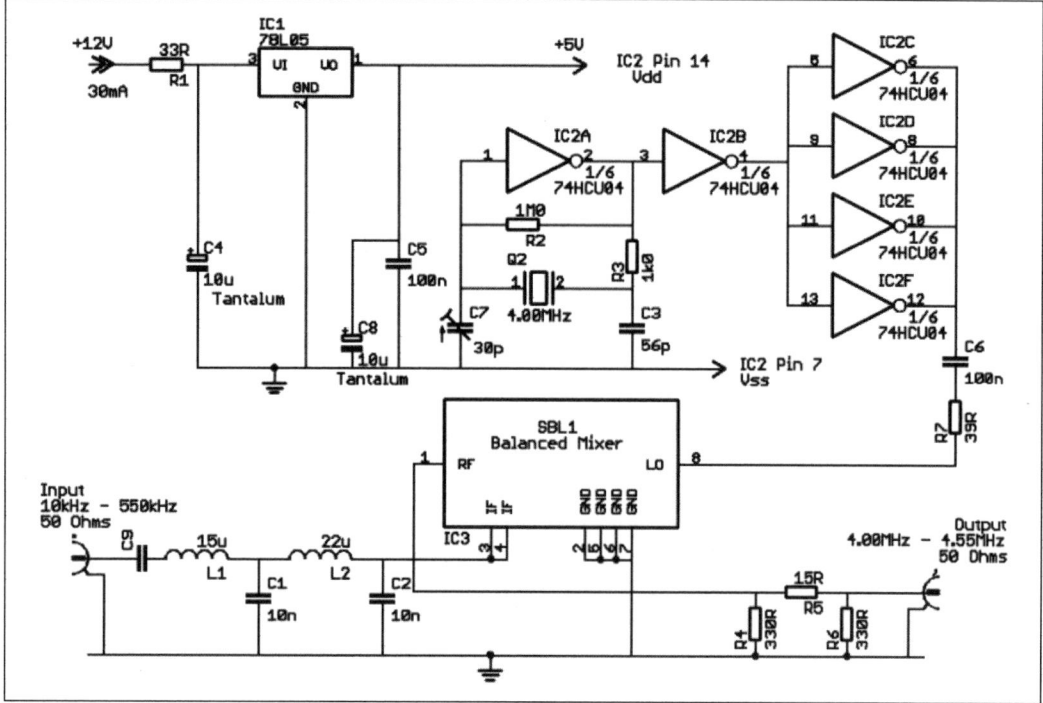

Fig 2.6: LF/MF - HF converter

and image responses, while frequencies much over 10MHz will lead to reduced frequency stability, which may be a problem when narrow-band modes such as QRSS are being received.

The converter output has a simple -3dB attenuator pad to reduce the effect of output impedance variations on the mixer; the circuit therefore has an overall loss of about 10dB. No post-mixer amplifier stage is included, since most HF receivers include a low-noise preamplifier that can be switched in to perform this function.

The crystal frequency can be adjusted by setting the HF receiver to exactly 4MHz (or other crystal frequency), and adjusting the trimmer capacitor so that the oscillator signal is exactly centred in the CW passband of the receiver. The received input frequency is then the value displayed by the receiver, minus the crystal frequency.

This converter gives good performance from 550kHz down to very low frequencies, and for receivers with very poor sensitivity at LF or MF will often give better results than the addition of high-gain preamplifiers. It is also an effective way of extending the lower frequency capability of HF-only receivers.

Commercially available add-ons

Low frequency converters in kit form or ready built can be obtained from the USA, for instance from WB9KZY [10] or Palomar [11], and others may be found second-hand on auction sites such as eBay. One commercial LF/MF transmitter made in Europe has a preamplifier and converter built-in (see the chapter on transmitters).

'SoftRock Lite' modifications for LF/MF

The SoftRock software-defined radio kits designed by Tony Parks, KB9YIG, have been popular as an entry-level SDR project for the HF bands. The same concept, with suitable modifications, is well suited to the narrow 136kHz and 472kHz allocations.

The modifications described here were originally applied by M0BMU to the SoftRock Lite v6.2 kit, and it included the 500kHz band. The Softrock kit has now been superseded, but these should be able to be modified in a similar way. **Fig 2.7** shows the modifications to the SoftRock circuit. The modification details have been changed from 500kHz to 472kHz.

Two major changes are required for 136kHz or 472kHz; the local oscillator frequency is changed, and the input bandpass filter is re-designed.

For 136kHz the crystal frequency frequency was changed to 1MHz. An alternative to the 1MHz crystal would be to use a 1MHz DIL oscillator module, with its logic-level output connected directly to the divider flip-flop. An easily obtainable 1.8432MHz crystal is suitable for 472kHz. With the appropriate frequency division jumper on the PCB fitted, this gave local oscillator frequencies of 125kHz and 460.75kHz respectively, resulting in tuning ranges of 101 - 149kHz and 436.75 - 484.75kHz with a 48kHz sample rate.

The existing SoftRock oscillator/driver circuit just works at 4MHz, but not 1MHz. Fig 2.7 shows changes to the circuit to obtain better oscillator waveforms at the lower frequencies.

The original SoftRock input filter could be re-designed for the LF/MF bands, but the components required would be too big for the tiny PCB, and the rejection of LF and MF broadcast stations near harmonics of the local oscillator frequency would probably not be good enough. Instead, the original tuned input transformer T1 is replaced by a wideband ferrite-cored transformer, and off-board filters were designed with increased rejection at harmonic frequencies. As in the original SoftRock design, component values are fixed, and no adjustment is required. Values are shown for each band. For the 472kHz band, small axial-leaded inductors the size of half-watt resistors were used and were quite satisfactory. For the 136kHz band, this type of component has insufficient Q, leading to high insertion loss and a poorly-defined filter passband. Instead, slightly larger radial-leaded chokes wound on ferrite bobbins (eg Panasonic ELC series [9]) were used. Their Q of around 40 was adequate.

Both LF and MF versions have been used, mainly with I2PHD's *Winrad* software [3] and DL4YHF's *Spectrum Lab* software [2], both with excellent results. If the connections to the PC sound card input are made according to the details on the SoftRock kit schematic, the 'reverse I and Q channels' option should be selected to give correct sideband selection. The phase and amplitude balance between I and Q channels should be adjusted to obtain maximum rejection at the

The tiny SoftRock version 6.2 receiver

CHAPTER 2: RECEIVERS

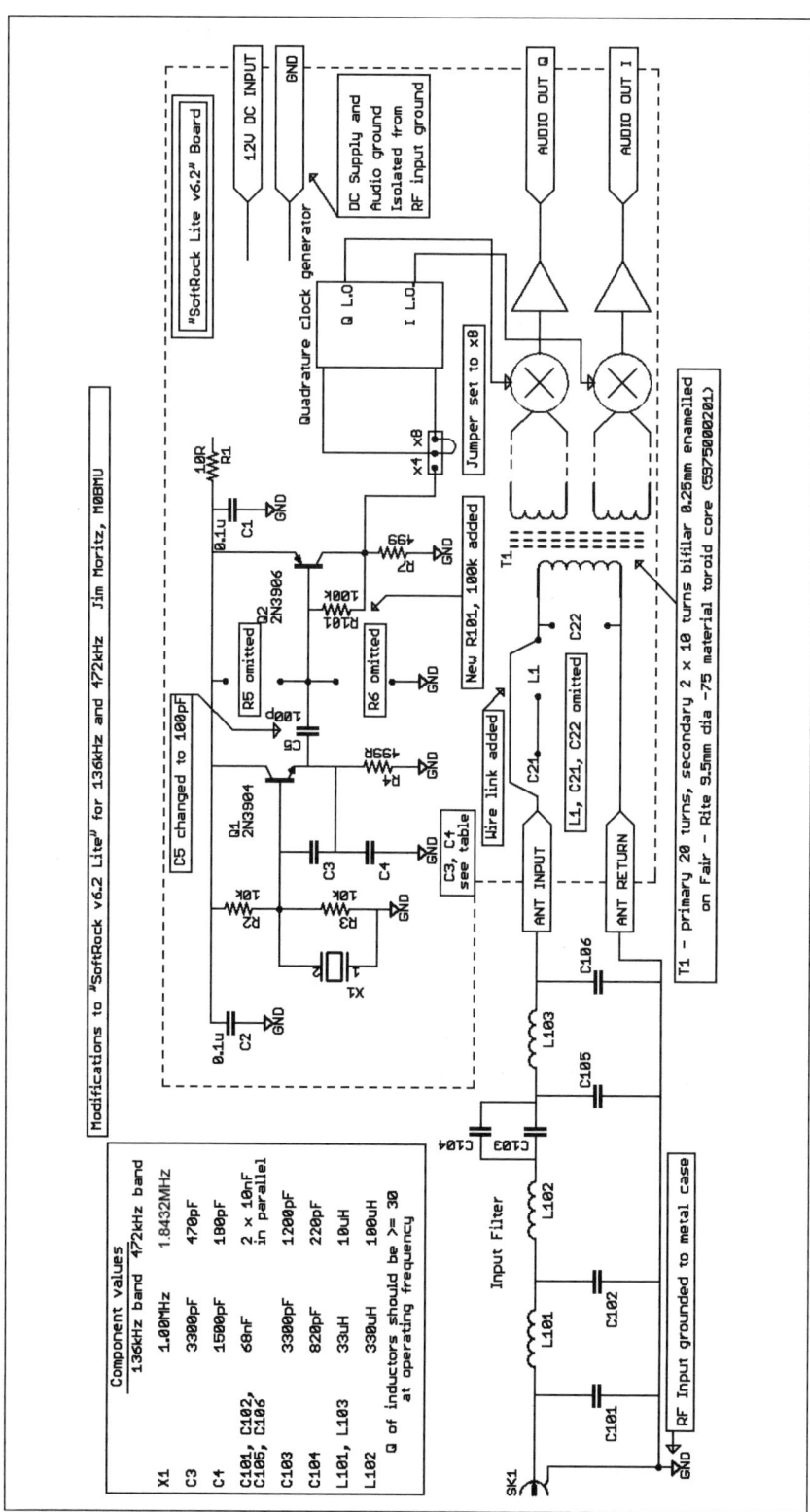

Fig 2.7: Modifications to SoftRock Lite v6.2 for 136kHz and 472kHz. Note that the circuit of recent versions of this kit may differ

'image' frequency of the centre of the amateur band. This is especially important on 136kHz, where strong signals are present around the image frequency. Sensitivity on both bands was around 1 microvolt for 10dB SNR with a CW bandwidth of 300Hz; this is not particularly high, but is quite adequate if a transmitting-type antenna is used, or a preamplifier or preselector such as the ones described earlier in this chapter.

In order to prevent low frequency noise generated by the PC and its power supply from getting into the sound card audio input, it was important to maintain isolation between the audio output ground and the RF input ground of the SoftRock circuit. The RF ground was connected to the metal case housing the SoftRock board and input filter, and the audio output and DC supply ground connections were insulated from the case. It was also necessary to use separate DC supplies for the SoftRock and preamplifier or transmitter to prevent ground loops. Connecting RF and audio grounds together within the case resulted in a 30 - 40dB increase in noise level.

The latest softrock boards can be purchased from [12] and documentation, including circuits, can be found at [13].

References

[1] 'Rocky' SDR software by VE3NEA at *http://www.dxatlas.com/Rocky/*
[2] 'Spectrum Lab' and other radio-related software can be downloaded via DL4YHF's web pages: *http://www.qsl.net/dl4yhf/spectral.html*
[3] 'Winrad' SDR software and other radio-related software can be downloaded from I2PHD's Weaksignals site: *http://www.weaksignals.com*
[4] SDR-IQ receiver: *http://www.rfspace.com/RFSPACE/SDR-IQ.html*
[5] Perseus SDR receiver: *http://microtelecom.it/perseus/*
[6] Afedri SDR-Net receiver: *http://www.afedri-sdr.com/*
[7] Selective Level Meter MV61 Used as a Receiver for VLF and LF *http://www.qru.de/MV61.htm*
[8] Frequency Selective Voltmeters and their Uses in the Radio Hobby *http://www.qru.de/selective%20level%20meters.html*
[9] RS Components on-line catalogue: *http://uk.rs-online.com/web/*
[10] LF Converter kit from WB9KZY: *http://wb9kzy.com/lfconv.htm*
[11] Palomar VLF Converter: *http://k1el.tripod.com/VLF.html*
[12] Softrock SDR kits: *http://fivedash.com/*
[13] Softrock SDR documentation and circuit: *http://www.wb5rvz.org/softrock_lite_ii/index?changeBands=no*

3

Antennas and matching

In this chapter:

- A tiny bit of theory
- Antenna options
- Marconi design considerations
- Danger high voltage!
- Using an existing antenna
- Ground systems
- Non-earth losses
- Purpose-built antennas
- Marconi loading and matching
- Transmitting loops
- Practical loop antennas
- Earth antennas
- Why use 50 ohms on LF?
- Antenna supports

THIS CHAPTER DEALS WITH the requirements of transmitting antennas, all of which can be used for reception. Antennas designed specifically for receiving are covered in the next chapter. Mathematics have, by and large, been kept to a minimum; the emphasis being on practical considerations. Those wishing to investigate the 'why' rather than the 'how' should read [1], [2] and [3].

As discussed earlier, the ideal antenna for the 136kHz band would be a vertical 550 metres high tuned against an extensive earth mat around 1km across. Even at 472kHz, you would need to use a 150m high tower. In practice, even commercial LF stations do not often use full sized antennas.

A compromise is reached by shortening the element length, bending it to reduce its height, and compensating for the 'lost wire' by adding a loading coil. As an alternative a capacitively tuned loop antenna may be used.

For a typical amateur installation, the degree of compromise is considerable. This gives rise to several problems, the greatest of which is a reduction in efficiency to a small fraction of 1% on 136kHz and only a few percent at 472kHz. Others include the need for a large loading coil and the presence - at least on 136kHz - of very high antenna voltages (or high currents in loop antennas) at quite low transmitter powers. Despite all that, it is possible to radiate a respectable signal from an antenna that will fit into a suburban garden.

A tiny bit of theory

The effective radiated power (P_{ERP} or P_{EIRP} - see the Measurements chapter) radiated by an antenna is determined by the product of its radiation resistance (r_{RAD}), the square of the antenna current (I), and the directivity (ie the gain resulting from the directional pattern of the antenna, shown as D):

$$P_{ERP} = r_{RAD} \times I^2 \times D$$

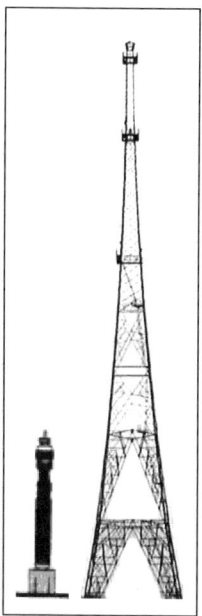

A full-sized vertical for 136kHz would be nearly three times as high as London's BT tower

> *See the chapter on Measurement and Calculations for the maths behind low frequency antennas, including how to calculate radiated power.*

We have control over two of these at LF: the radiation resistance of the antenna and the current passing through it. The radiation resistance is related to the antenna's size and shape, whilst the current is a function of the applied RF power and the total resistance. Unfortunately, the radiation resistance is not the only resistance consuming the transmitter power, there are also the loss resistances. These losses occur within the antenna and its matching system, and in the environment of the antenna (the ground system, objects near the antenna). On HF these loss resistances are often negligible as they are small compared to the radiation resistance, but at LF this is certainly not the case.

For most amateur antennas the radiation resistance on the 136kHz band is in the range of one to a few hundred milliohms (yes, thousandths of an ohm!), while loss resistances are in the range of 30 to 200 ohms. This means that, dependent on the antenna and its environment, about 99% to 99.99% of the transmitter power is not radiated but absorbed in the loss resistances. This is why it is actually quite difficult for the average amateur station to reach the UK licence limit of one watt effective radiated power.

Typical radiation resistance rises to somewhere between a few tens of milliohms and an ohm or two at 472kHz and losses fall to about half of the 136kHz figure. This results in an efficiency improvement of 20-30 times that on the lower band. Because of this, it is relatively easy to approach the UK's 5W EIRP limit.

The formula for calculating the radiation resistance of a monopole is given in the measurements chapter. Formulas for caculating antenna efficiency and hence radiated power can also be found in that chapter.

At a distance, the electric and magnetic fields produced by any antenna are both the same for a given effective radiated power. Close to a loop antenna, however, the magnetic field is relatively large, while the electric field near a vertical monopole predominates.

Antenna options

Those with very large gardens may think in terms of a loaded horizontal dipole for 136kHz. Unfortunately, the effective height would be so low as to render such an antenna totally ineffective. There are two practical options: a loop or a Marconi (vertical) (**Fig 3.1**). Experiments by UK amateurs have shown the Marconi to have some advantages, particularly on 136kHz, but the loop continues to have its devotees, especially in the USA.

An LF transmitting loop comprises a single turn of low-loss wire (there is no advantage in a multi-turn system) with as large an area as possible, with its lowest point a metre or so above ground (**Fig 3.1c**). The loop must be tuned to resonance and matched to the feeder or transmitter. Its chief advantage is that, as described above, it is much less affected by lossy nearby objects, eg trees and houses, than a Marconi and it does not require a good ground. The loop has a figure-of-eight directional pattern; this is a disadvantage for transmission due to the deep nulls in the transmitted signal at right angles to the loop.

The classic Marconi is a vertical monopole with the top part bent over to form a capacity hat. Typical designs for HF and MF use are the 'T' and the inverted 'L'. These work fine at LF, though any shape of top section will work so long as it has sufficient capacity to ground. An efficient ground system must be provided.

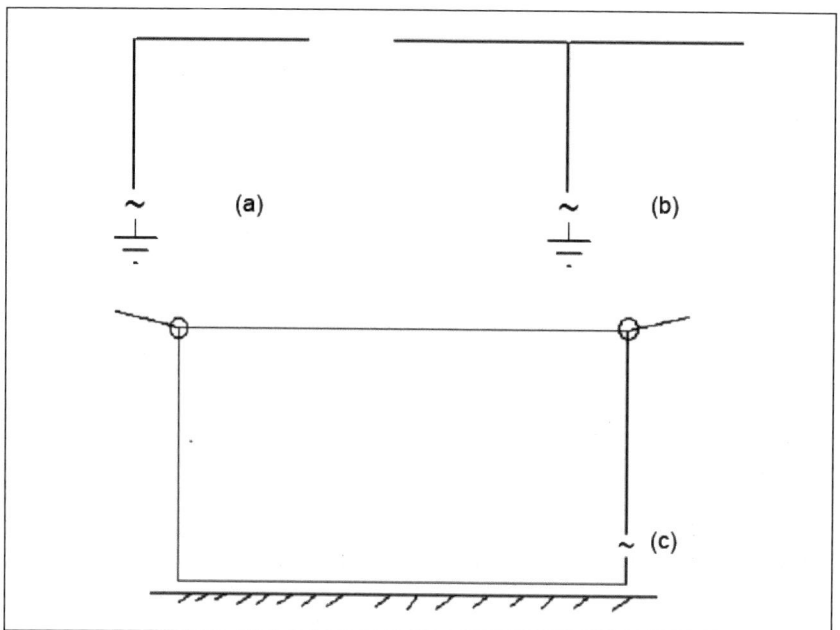

Fig 3.1: LF transmitting antenna options: (a) inverted-L Marconi, (b) Marconi tee and (c) loop.

Design considerations of an LF/MF Marconi

The formula for calculating radiated power is dealt with in detail in the measurements chapter. For the purposes of the design of a vertical antenna it is important to note that it includes on the top line the term h_{eff}^2. This means that the radiated power is directly proportional to the square of the effective height of the antenna. So doubling the effective height will increase your radiated power by four times, and any increase at all in h_{eff} will be useful.

So what is effective height and how does it differ from the actual height? If you consider the current distribution on a short vertical (and all amateur low frequency antennas can be considered to be short), it can be seen that the current is at a maximum at the bottom - usually the feedpoint - reducing approximately linearly to zero at the top (**Fig 3.2**). The average current is half of the maximum current. The effective height is therefore considered to be half of the total height.

If a horizontal capacity hat is added, as in the inverted-L, the effective height will be increased from this 50% figure, by an amount determined by the size of the capacitance. Note that the effective height will always be between 50% and 100% of the height of the vertical section. A very large amount of horizontal wire will bring the effective height close to 100%. Any part of the capacity hat that is lower than the top of the vertical section will tend to reduce the effective height.

A bonus from having a substantial capacity hat is that it reduces the amount of inductive loading need-

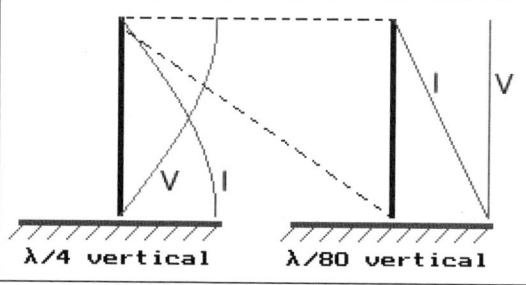

Fig 3.2: Current and voltage distribution on a full-sized, and very short, Marconi antenna

Fig 3.3: Multiple wire top loading

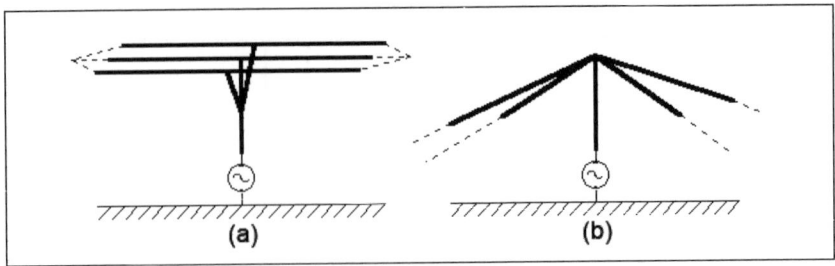

ed to bring the antenna to resonance, ie the loading coil can have fewer turns and hence less loss.

The vertical section should be as vertical as possible because any horizontal component will have a capacity to ground which will detract from the effective height. There is no need for multiple wires in the vertical section. The vertical wire should also be as far away as possible from other vertical items such as house wiring and trees.

It is often necessary to run the vertical wire close to the pole or tower that supports it. In this case, unless the pole is very well insulated from ground and other supports (walls, guy wires), losses can be reduced by grounding it. You may think that this will 'short out' the vertical wire, but in fact it simply provides a low-loss capacitance to earth instead of dissipating power in the lossy return to earth. If in doubt, try it and see for yourself.

In general terms, the capacity hat should comprise as many wires as possible, as far apart as possible and covering as much ground as possible.

Without getting into too much detail, the key physical factor is that there is a logarithmic variation of capacitance with both diameter of wire and height of wire above ground level. A calculation for doubling the wire diameter from 1mm to 2mm results in an increase of capacitance of only some 7%. The height above ground also only has a minor effect on capacitance (though see above regarding effective height). The practical implication is that the wire diameter used and the height above ground are not critical factors in determining capacitance. The length of the top loading wire is the main factor.

Results of capacitance calculations based on reference values of 1mm diameter wire and a T antenna with a height of 10m above ground are:

- vertical downlead approximately 6 picofarads per metre of wire
- top loading approximately 5 picofarads per metre of wire

The Golden Rule of short Marconis:

Get as much wire in the air as you can, as high as possible

We shall use these figures in a later chapter to calculate the approximate radiated power from a Marconi antenna.

Further calculations have been carried out to explore the impact of running two same diameter parallel wires, rather than a single wire. The reference condition again being wire of 1mm diameter at 10 metres above ground. The following results were obtained:

- Two wires separated by 1 mm increases capacitance by about 4%
- Two wires separated by 10 mm increases capacitance by about 19%
- Two wires separated by 100 mm increases capacitance by about 39%

- Two wires separated by 1 metre increases capacitance by about 68%
- Two widely separated wires increases capacitance by up to 100%, ie double the capacitance of a single wire, as one would expect.

The results indicate how proximity effect limits the realisable net capacitance for closer spaced wires. However, despite proximity effect, it is generally the case that two wires are better than one when it comes to increasing the net capacitance. Note that doubling the wire diameter results in four times the surface area and happens to result in 7% more capacitance, whereas two same diameter wires that are nearly touching involve two times the surface area and happen to have a 4% increase in capacitance. Note also that doubling the wire diameter gives a weight increase of four times, so in terms of mechanical support and sag of wires in an antenna, two separate wires again are generally better than a single thicker wire. It is clear that multiple wires are the answer to arranging useful top loading. It is also clear that this is already known to the designers of commercial LF beacon antennas, as T antennas are extensively used, with typically two or three parallel wires in the top loading, with spacing of a metre or so (**Fig 3.3(a)**). In practice, the wires need not be the same length or in a symmetrical pattern.

It is tempting to increase capacitance by bringing the ends of the capacity hat down towards the ground (**Fig 3.3(b)**) as is done with full-sized HF antennas such as the inverted-vee. However, the effective height, and therefore the efficiency, of a short low frequency antenna is considerably reduced if the capacity hat droops too much.

The wires should not run too close to lossy objects such as trees and buildings. It can be useful to keep trees and large shrubs pruned regularly.

Danger, high voltage!

Extremely high voltages - tens of kilovolts - can be expected on an LF Marconi tuned to 136kHz, even with transmitter powers of a few tens of watts. For this reason, never allow any part of an LF antenna to touch trees, walls, window frames or even plastic guttering. Not only will this cause large losses (ON7YD had a 14% reduction in antenna current when a single leaf touched his antenna), but there is also a significant risk of fire. Note that, contrary to what you might think, a small antenna will have higher voltages on it than a large one. To calculate the voltage, use the formula:

$$V_{ant} = I_{ant} \times \frac{1}{2\pi f C_A}$$

At 472kHz, the smaller power levels and greater antenna efficiency reduce voltages to a few hundred volts, but care must still be taken.

Very high antenna voltages can produce an effect known as 'corona discharge' in which tiny sparks emit from wires, even through insulation. Corona eats away at plastic items such as cable ties. It is tempting to rely on nylon rope as an end-insulator, but this will often lead to the rope melting and the wire falling down. Instead, use good quality insulators. If two insulators are used in series, join them with short lengths of wire, not plastic. Even after taking these precautions, it is important to reduce the corona because the sparks are a form of power loss. Corona discharge is far worse at sharp points, such as the ends of wires or right angle bends. Losses can be reduced by ensuring that all bends are gradual. Wire

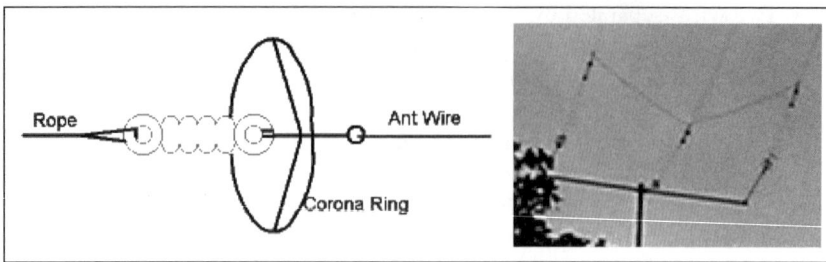

Fig 3.4: Two ways of reducing corona discharge from wire ends: (left) A ring termination, and (right) ends joined together. The photograph shows the far end of a three-wire capacity hat. Note the two sets of insulators on each wire

ends can be protected by forming them into a small loop. Corona on a multi-wire top section can be reduced by joining the ends together (**Fig 3.4**). Corona is almost invisible, but it may be possible to detect it on an FM broadcast radio.

Precautions against high voltage flash-over should extend to all parts of the antenna system, including the loading coil and any tuning mechanism. It is especially important to pay attention to any part of the antenna that passes through the window or wall of a shed or the house.

A common problem is how to run an LF antenna wire through a window frame from the 'shack' to the outside world. Often the loading coil is located in the shack for ease of tuning and to keep it out of the weather. The voltage on the antenna system is far higher after the loading coil, so a preferable arrangement is to mount the coil in a weatherproof box such as a plastic dustbin outside the shack. The wire between the coil and the transmitter then has a much reduced voltage on it. Alternatively the entire loading, matching and tuning system can be located outside and fed with coaxial cable from the shack.

Any part of the antenna system that can be touched should have clear warning signs, but this is no substitute for making it inaccessible to humans and animals.

Using an existing antenna

With care, an existing MF, HF (or even VHF) antenna installation can be pressed into service as an 136kHz or 472kHz antenna, usually by strapping the feeder to form a vertical element and using the MF/HF antenna as a capacity hat. The system should, of course be tuned against a ground system (see later).

An ideal set-up would be a dipole for 1.8MHz, with the shack directly underneath its centre. However, smaller antennas will work, especially if they are at a good height. It is possible to modify feeder runs when the shack is not in the ideal position.

As described above, the vertical section - in this case the dipole's coaxial feeder - should be predominantly vertical and clear of any nearby objects. It may well be necessary to divide your coax run into the vertical bit that goes to the antenna and the horizontal bit that runs from the shack. Adding a connector at this point will make it simple to change from the Marconi configuration back to the HF arrangement

A box should be installed at the foot of the vertical section, to contain the loading coil and any matching device. This will need an enclosure at least 0.5m cube, but preferably larger. A dustbin or small shed is ideal. In the intense RF fields near the loading coil and antenna feeder, objects such as plants, trees or masonry may cause significant RF losses. If the loading coil enclosure cannot be positioned away from these, an enclosure lined with (or made of) metal

CHAPTER 3: ANTENNAS AND MATCHING

which is earthed can improve efficiency by screening the coil from the lossy materials.

Do not be tempted to install a switch to change the antenna configuration from HF to LF. The low frequency antenna is likely to have very high voltages on it when transmitting and these can cause flash-overs that may damage your HF radio. It is preferable to do the change-over by moving coax plugs (**Fig 3.5**).

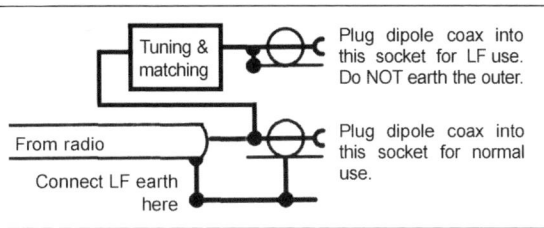

Fig 3.5: Simple method of switching a dipole for 160, 80 or 40m between use as a dipole and use as a Marconi for 136kHz and/or 472kHz, by using coax sockets

The horizontal section of an HF dipole may already be suitable if it is well engineered, but it is important to check that it does not droop too much, that it does not touch or even run too close to trees or walls, and that the insulators are of high quality. Making improvements in these areas may also boost the performance of the antenna when used on the bands they were originally intended for.

An 160m or HF ground-plane vertical or inverted-L can be pressed into service on the low frequencies using a similar technique.

An entirely horizontal wire will not be effective at LF. However, if a long-wire antenna can be contrived to have some vertical component, perhaps by introducing a slope or moving the feedpoint to ground level, it can work surprisingly well.

It may be possible to use an HF, or even VHF, beam antenna if the mast is well insulated from the ground, or the tower can be used as the support for a sloper for LF.

Ground systems

A vital part of a Marconi antenna is the ground system, which forms the return path for the antenna current to the transmitter output. The ground system can be as simple as a single earth rod driven into the ground, or as complex as hundreds of radial wires. Commercial low frequency stations use hundreds of kilometres of wire spread out over a radius of several hundred metres.

How easy it is to provide an efficient ground connection, and indeed the efficiency of the antenna itself, will depend on the conductivity of the soil beneath the antenna and extending for some kilometres away from the station. Sea water is ideal and proximity to the sea can be a great advantage. Clay can be useful, whilst sand and rock tend to be poor. Amateur LF stations have been operated from all types of terrain, so do not be discouraged. Poorer soils just need a bit more work and some ingenuity.

The most basic earth connection is a single metal stake driven a metre or so into the ground. The current density is highest at the bottom of the vertical section, so the earth stake should be as close as possible to this point. This simple arrangement will get your station on the air, but should not be regarded as optimum.

In practice, successful LF earth systems include a mixture of several earth stakes, radials and any other earthed items such as water pipes and even tin roofs.

Earth stakes should be a metre or two long. Copper water piping is cheap and easily obtainable but it is soft and will eventually bend rather than go any deeper. Purpose-designed copper coated steel earth rods are now available from big hardware stores.

29

> **WARNING**
> Before driving an earth stake into the ground, it is vital to make sure you avoid any buried services, such as water, drains, telephone, cable TV, gas or electricity.

When using a Marconi antenna on 1.8MHz, radial wires, sometimes raised above ground, are often used in preference to earth stakes. This is still an option on lower frequencies, but any radials will be a much smaller fraction of a wavelength and hence much less efficient. Radials are likely to be much more effective on 472kHz than on the lower band. The wires should be as long as possible and can be run along the surface of the ground, or just below it. There seems little agreement on whether the wire should be bare, ie in direct contact with the soil, or insulated.

As a general rule of thumb, a low frequency earth system should comprise as many elements as possible, but this is not the entire answer. Firstly, the efficiency of the earth system is not directly proportional to the number of elements involved, and the results obtained from the second set of radials will be rather less than the first set, and so on. A point will be reached where adding further elements to the ground system produces little or no improvement. Increasing efficiency further will then require improvements to other parts of the antenna system.

Secondly, it has been found that not every additional earthing element improves the situation. This may be because the reactances inherent in each element act against each other. So when building your earth system, it is useful to bring each separate element to a connection strip, so that the effect of each one, or a combination of elements can be tested.

Measuring the RF current flowing into each earthing element will show which are the most effective. Suitable current meters are described in the measurements chapter.

Although a good earth is important for a Marconi antenna, do not spend too much effort on it because the other losses in the system can be much higher. Time and energy are better spent getting more wire in the air, and higher.

> **WARNING**
> Many modern houses in the UK have an electricity safety earth system known as Protective Multiple Earthing (PME). Under some fault circumstances this could pose a risk to amateur stations. It is recommended that your RF earth is bonded to the PME bonding point using cable of at least 10mm^2 in accordance with IEE wiring regulations. A leaflet on this subject, endorsed by the Electricity Association, is available from the RSGB's EMC Committee.

Non-earth losses

Experiments by M0BMU show that a particular inverted-L antenna erected in a domestic back garden surrounded by trees had a loss resistance measured at 61 ohms on 136kHz and 26 ohms on 502kHz, while a near identical antenna erected in an open field for comparison had loss resistances of only 8.5 ohms at both frequencies. These losses (**Fig 3.6**) are often lumped into "earth resistance" in calculations and lead to the assumption that losses in the earth system itself are the most critical factor. It is more helpful to describe these as environmental losses. These losses come from absorption of the signal by trees, shrubs, buildings etc, and from the screening effect caused by them. This can significantly reduce the effective height of an antenna and therefore the radiated power.

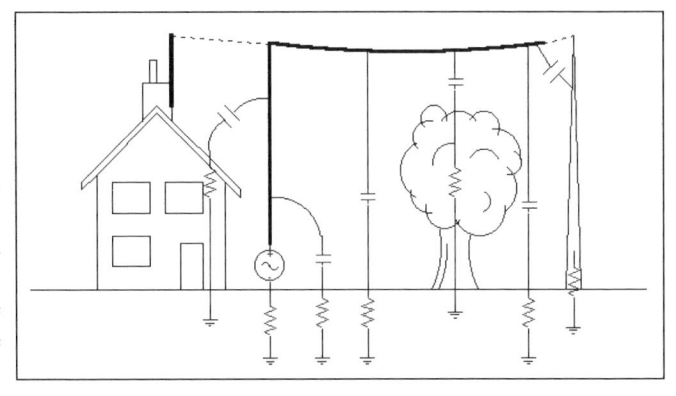

Fig 3.6: Loss in the environment around an antenna

Purpose-built antennas for small gardens

When designing a low frequency antenna from scratch, there are two important considerations: height and ground coverage.

Ideally the garden should be clear of RF absorbers such as trees and shrubs, the antenna should be well away from the house and you should have several poles or towers around the periphery of the garden, plus one in the centre supporting the vertical wire. However, in practice, every system will involve some kind of compromise, often a great deal of compromise.

Try not to compromise too much on height. The effective height has a square law influence on the radiated power, so try to make the vertical section as high as possible whilst avoiding too much droop on the horizontal sections.

The top section, or capacity hat, of a Marconi antenna can be any shape, but the most popular and practical are the L and T. The key to success is to cover as much new ground as possible but the arrangement chosen will depend on the environment, ie how many high supports exist or can be erected, how much real estate is available and, importantly, what the neighbours will tolerate. Remember that any new antenna system may require planning permission.

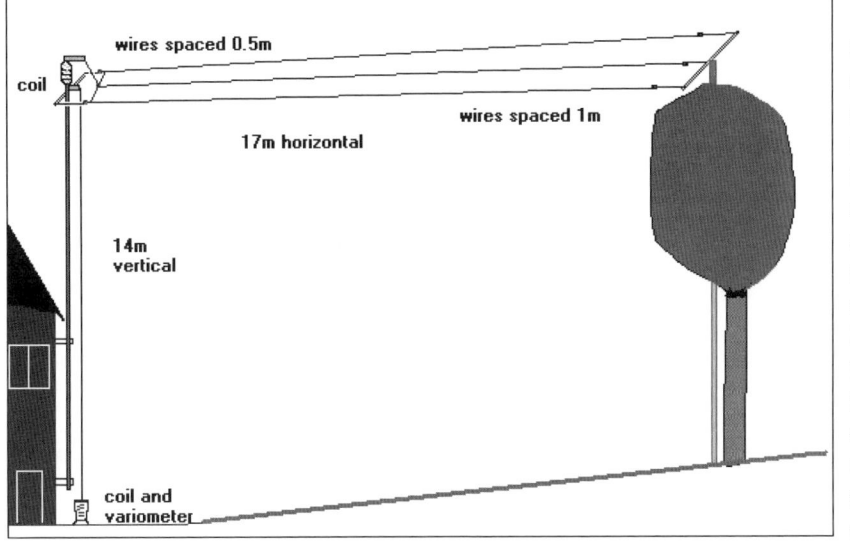

Fig 3.7: An early suburban LF antenna system at G3XDV, using a house-mounted mast, and another strapped to a tall tree. The antenna uses both top and bottom inductive loading plus three top wires. This set-up was used to make a 136kHz transatlantic contact

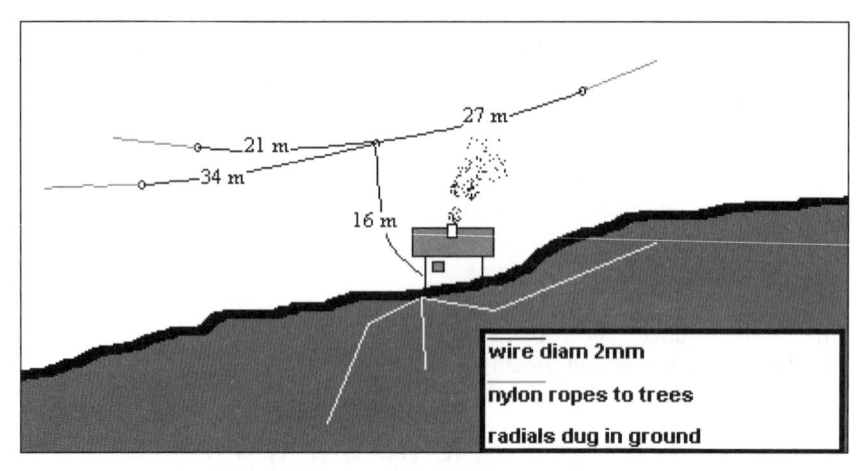

LF TODAY

Fig 3.8: A larger set-up at OK1FIG's summer house has a top section of different lengths of wire going to a variety of handy supports. This arrangement led OK1FIG being heard at world-record distances

(below) Fig 3.9: An umbrella antenna has several short top wires bent downwards to make a capacity hat. This illustration comes from [1] where a formula can be found for optimising this type of antenna

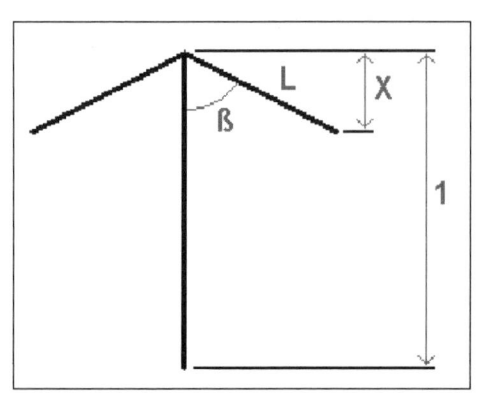

(right) Fig 3.10: A spiral top has elements of inductive as well as capacitive loading

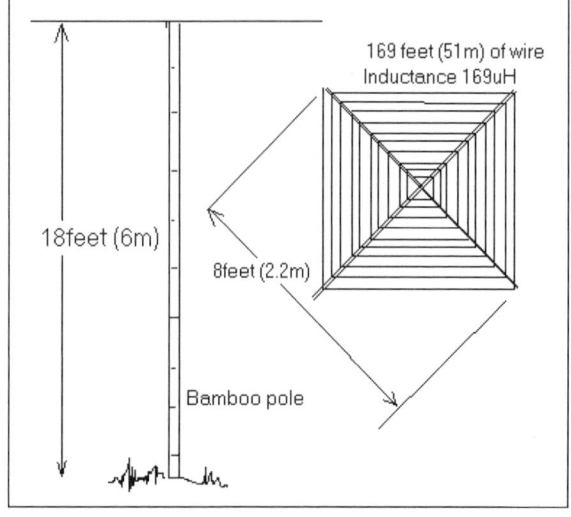

A multi-wire top section may use parallel wires in the classic L or T configuration (**Fig 3.7**), or simply as much wire as can be strung between tall trees on the site (**Fig 3.8**). The wires should be as far apart as possible; the closer they are, the less benefit is achieved. It can be useful to join the ends of multiple wires. This may add a little to the capacitance, particularly if the wires are spaced far apart, but will mainly reduce corona discharge. Joining the ends may also help to keep most of the antenna aloft if one of the supports breaks.

Another capacity hat configuration used by commercial LF stations such as non-directional beacons (NDBs) is the umbrella (**Fig 3.9**). Its main advantage is that it requires only one mast, and the top wires can be part of the guying arrangement. Although this is less efficient because it reduces the effective height, it may be possible to compensate by increasing the RF power.

A spiral top section (**Fig 3.10**) has been tried with some success as it combines capacitance with some extra inductance, but it is quite difficult to engineer.

Are more short wires preferable to one longer wire? This comes back to the aim of as much capacitance as possible without sacrificing too much height. A single very long top wire can be extremely effective, especially if it is at a good height (15m or more), but a similar effect can be achieved by many shorter wires. It is really a matter of what is the most convenient. If you have the resources for one long and many short wires - go for them all!

If you have the option of planning where to put the vertical section, rather than this being dictated by circumstances, it should be as far as possible from other vertical structures, including trees, masts and house wiring. The top wires must not touch any part of a tree - this will lead to losses, unpredictable performance and (on 136kHz) fires. Instead, keep all wires a few metres away from trees if at all possible. Similarly, avoid running wires above tall shrubs - or cut the shrubs first.

Loading and matching a Marconi

Any practical low bands vertical will be very much shorter than the length needed to be self-resonant. The resultant capacitive reactance must be compensated for by the addition of a series inductance, known as a loading coil. Depending on the antenna's size, and the frequency in use, this inductance can range from 0.1mH to more than 4mH. In addition, some kind of tuning mechanism will be required.

For receiving, almost any coil of the appropriate inductance will serve to resonate the antenna. However, a coil used for transmitting must be able to withstand the high current and voltage involved, as well as have the lowest possible losses. It must also be kept out of the rain.

The loading coil is a critical part of any LF antenna system. It must be home-constructed because nothing suitable is available off the shelf. For this reason, there are as many different types of loading coil as there are LF stations.

To start with, a coil former must be found. This normally entails a visit to the local DiY store. The former must be made of a material with low RF loss - if in doubt, cut off a small piece and cook it in a microwave oven, together with a cup of water (important). If the material gets hot, don't use it. In general, expect lighter coloured material to be less of a problem than darker.

The coil former is typically between 100 and 500mm in diameter and 200-500mm long. The actual dimensions will vary, depending on the thickness of the wire used and the inductance required. The sides of the former must be parallel

Two examples of loading coils for 136kHz: (left) a thing of beauty from CT1DRP is wound with aluminium wire and the integral variometer can be clearly seen, whilst (right) a wheelie-bin coil, tapped every ten turns proved useful for the DF2BC/P lighthouse expedition

LF TODAY

YU7AR's variometer tuned inductor

> **Should I use Litz wire?**
> Because the AC current flows only on the outer part of the wire in the coil, RF losses are greater than the DC resistance (the skin effect). A typical 136kHz inductor may have RF losses of 10 ohms or more, which in a well engineered system may be significant, so it can pay to use Litz wire. This is specially designed multi-stranded wire that has each of its wires insulated from the other. It is quite expensive and difficult to obtain, but worth grabbing if some becomes available. However, it is vital to strip the enamel insulation from every wire, or the losses will be increased (see also Appendix 2).
> Because of the smaller value of inductance required for operation on 472kHz, there is less advantage to be obtained by using Litz wire.

- don't be tempted to use something tapering like a bucket as the turns will eventually move. It is not essential for the former to have a circular cross-section.

Formers that have been successfully used include drain piping (or several pipes), rolled up plastic fencing, compost bins, fizzy drink bottles and plastic boxes. The variety is illustrated on these pages.

Losses in the loading coil reduce the power that can be radiated, so care must be taken in construction. You should also take into account the fact that the coil may have several tens of kilovolts on it, especially on 136kHz.

The 'goodness' of an inductor is called its 'Q'. A typical LF loading coil will have an *unloaded* Q of 100 to 400, depending on the construction and wire type used. Factors affecting the Q include the coil's size and shape, the type of wire used and the spacing between the wires. The optimum shape is a width to diameter ratio of 2.5, although a more commonly used ratio is 1 or a bit less, probably for practical constructional reasons. The optimum spacing between wires is the width of one wire. In practice, this can be approximated to by using plastic covered wire, so that the insulation on adjacent wires keeps them about one wire apart. Wire of about 1mm to 1.5mm diameter is popular. Enamelled wire can be used, so long as the insulation is thick enough to withstand the voltage between adjacent turns.

Some stations use inductors wound with Litz wire to reduce losses (see box above), particularly on 136kHz, but for most

GW4ALG's variometer was remotely tuned by this belt drive

installations it is sufficient to use plastic covered flex with a low DC resistance.

It may be that an inefficient system, which has significant earth losses, may not benefit immediately from a few ohms reduction in inductor loss. However, the golden rule with developing your low frequency station (or any other for that matter) is to reduce losses whenever you can, an ohm at a time, improving your signal a fraction of a decibel at a time - the fractions will all eventually add up to a big signal.

On the HF bands, it is normal to use a combination of inductors and capacitors in an ATU to tune out any reactance. Some 472kHz operators have used this technique successfully with very low power transmitters, but at powers greater than a watt or two the voltages on the coil are so large that any capacitor is likely to flash over.

Under most practical conditions, any tuning must be carried out in the inductance itself - a variable inductor is required. A crude, but simple, way of producing a variable inductor is to make 'taps' on the coil every few turns. This is certainly an excellent method if you have little idea of the inductance required to tune the antenna.

Loading coil, variometer and matching transformer built by GM4SLV for MF use. It is all mounted in a five gallon beermaking bin

Fine tuning can be done by sliding ferrite material in and out of the inductor. Beware, though, that ferrites vary considerably and many get very hot in this application, with attendant losses. Broadcast radio ferrite rods are often unsuitable for 136kHz, whilst pieces of ferrite from demolished switch-mode power supplies can be effective. Take care not to put your fingers anywhere near a loading coil carrying RF - use a long plastic rod to insert the ferrite.

A better and more conventional method of varying inductance is to make another, smaller coil that rotates inside the larger one. Depending on its position, a portion of the inductance of the small coil either adds or subtracts from that of the large coil. This is known as a variometer.

The inductors described below have a wide range of inductance and are ideal for experimental purpose, such as when a new station or new antenna is being set up, or for portable use. The basic principles can be used when graduating to a system that requires much less adjustment, such as that used in an established station. It is well worth keeping the multi-purpose inductor, however, for any future experiments or for giving to someone new to the band.

When the antenna has been finalised, another coil can be constructed with a fixed inductance slightly less than is needed. Then a smaller inductance, with a variometer coil, can be placed in series with the large one and used for fine tuning within the amateur band. It is sometimes convenient to place the large inductor outside, beneath the antenna, and to house the variometer coil in the shack. Alternatively, a remotely tuned variometer may be constructed.

You may wonder why an antenna needs fine tuning in a band that is only the width of an SSB transmission. An efficient low frequency antenna is likely to have a high enough Q to have a 3dB bandwidth of only 1kHz or so. Furthermore, the antenna's capacitance will change with the seasons and the weather, due to changing losses in the soil and adjacent objects such as trees.

Spreading the inductance

It is possible to make an improvement to a short Marconi, ie one with relatively little top capacitance, by moving most of the inductance to the top of the vertical section (**Fig 3.11**). This reduces the voltage on the downlead, and in turn reduces the losses to nearby objects such as trees and houses. Theoretically an improvement of about 3dB can be achieved. However, because there is less of the antenna beyond the elevated coil, more inductance is required with consequent resistive losses. Mounting the coil may also present difficulties. Elevated inductors have been found to have a diminishing effect with increasing capacity hat size.

A light weight inductor is shown below. It is wound on a two litre fizzy drink bottle. To make the bottle firm enough to wind the coil, first remove the screw cap, then place it in a freezer for a few hours. Remove it from the freezer and immediately replace the cap. The cold air expands to make the bottle feel solid.

An alternative which has been used successfully, is to spread the inductance along the entire length of the antenna by helically winding it. The engineering problems have not made this a popular option.

Fig 3.11: ON7YD's antenna uses four trees as supports. He reduces losses in the trees by mounting most of the loading coil at the top of the vertical section

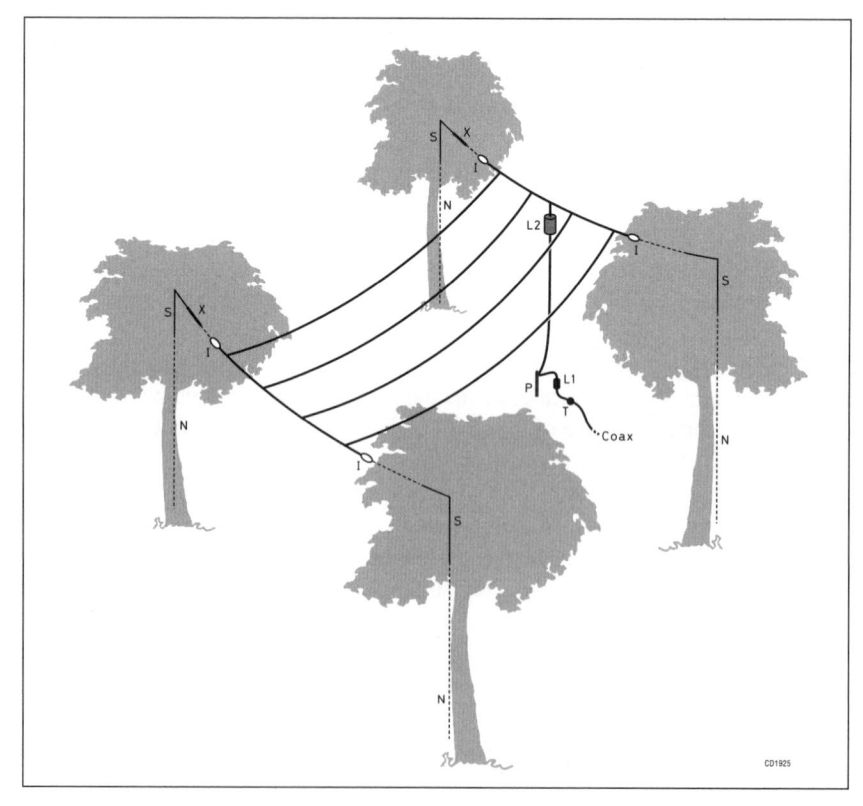

A pressurised drinks bottle makes a very light-weight inductor for mounting on top of the mast

Matching Vertical Antennas
In principle, you could use the same matching networks used for HF antenna matching, such as the pi- or T networks, to match low frequency. But in practice it is found that the component values, particularly for capacitors, are impracticably large, and for 136kHz require very high ratings due to the high antenna voltage. The two most popular LF/MF antenna matching circuits are shown in **Fig 3.12**.

In **Fig 3.12(a)**, a series loading coil has an inductive reactance that cancels out the capacitance, C_{ant}, of the antenna. The resistive component of the impedance (practically equal to the loss resistance) is then matched to 50 ohms, or other value of transmitter output impedance, using a ferrite-cored transformer. The capacitance of back garden amateur antennas, typically hundreds of picofarads, corresponds to a loading inductance of a few millihenries at 136kHz, and a few hundred microhenries at 472kHz. For most antennas, R_{loss} is between perhaps 10 and 200 ohms, requiring transformer turns ratios between about 1:2 and 2:1 to match to 50 ohms.

One design of 136kHz matching transformer, satisfactory at power levels up to 1kW, uses an ETD49 transformer core in 3C90 ferrite material, wound with 32 turns of 1.5mm enamelled copper wire, tapped every two turns. A much smaller transformer can be used at 472kHz. For power levels up to 20 watts or so, a Fair-Rite 22mm diameter, 43 material toroid core (5943007601) wound with 25 turns of 0.6mm wire, and again with taps every two turns, is suitable. The 50-ohm transmitter output is connected at the 16 turn tapping point, and the 'cold' end of the loading coil connected to the tap that gives optimum matching.

Because the antenna reactance is much larger than the resistance, the loading inductor must be capable of fine adjustment to obtain resonance accurately. As described above, coarse tuning is achieved by changing the coil tapping point, and a variometer allows fine tuning over a narrow range. This matching arrangement

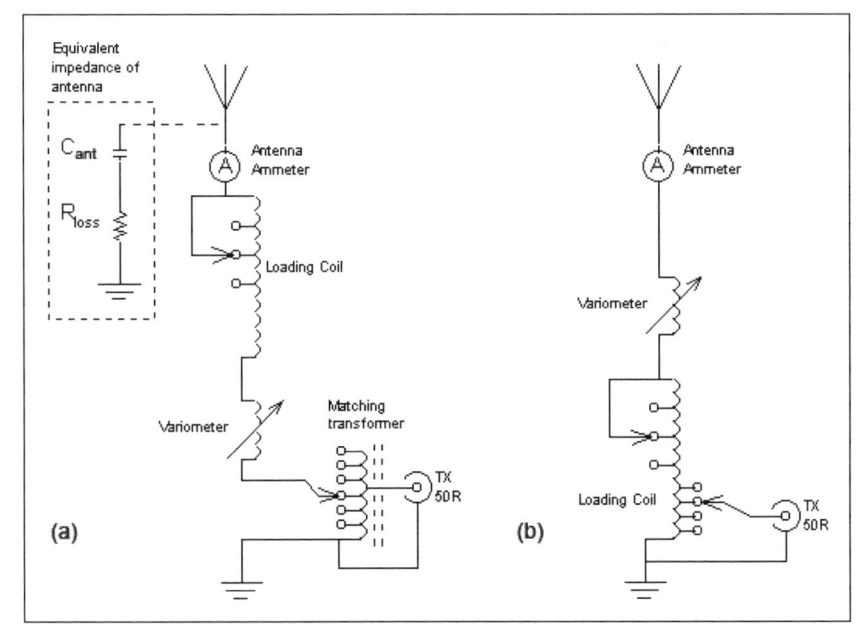

Fig 3.12: (a) LF antenna tuner; (b) Alternative antenna tuner circuit

(above) Antenna tuners at M0BMU using the circuit in Fig 3.12 (a) - 136kHz (left), 472kHz (right)

(right) OK1FIG's multi-tapped loading coil is used for tuning as well as matching as in Fig 13.3(b)

is very straightforward to use, since the adjustment of antenna resonance and resistance loading are almost completely independent. More information on using ferrite materials at low frequencies can be found in Appendix 2.

Another popular matching network uses a tapped loading coil as shown in **Fig 3.12(b)**. The low potential end of the coil is equipped with closely-spaced taps, so the loading coil also performs the function of the matching transformer. Although this is physically simpler than Fig 3.13(a), the electrical behaviour of this circuit is more complicated.

The primary and secondary of the transformer are not tightly coupled, so the transformer impedance ratio will not closely correspond to the turns ratio and the adjustment of the antenna to resonance and selection of the impedance-matching tap will be somewhat interdependent. However, it is not difficult to find a suitable tapping point by trial and error, and this will not often then need to be changed.

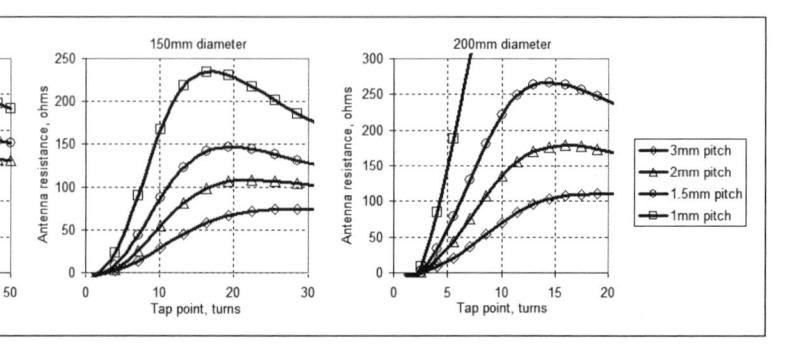

Fig 3.13: Antenna loss resistance matching with tapped loading coils of different diameter and winding pitch

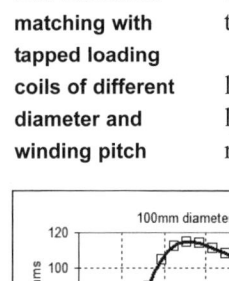

The range of antenna loss resistance that can be matched using the tapped loading coil depends on the coil geometry. In general, if the coil has a relatively small diameter and coarse winding pitch (ie it is wound with thick wire), the maximum value of R_{loss} that can be matched is quite low. If the coil has large diameter and fine winding pitch, much higher R_{loss} can be matched. However, the coil is then less suitable for low resistance antennas, because the required tap is only a few turns from the grounded end of the coil, giving very coarse steps in matching adjustment.

The graphs in **Fig 3.13** show examples of the range of antenna loss resistance that can be matched with coil diameters of 100mm, 150mm and 200mm, and various winding pitches. A Microsoft Excel spreadsheet is available for calculating similar curves for any coil diameter and winding pitch [5]

Sometimes it is desirable to isolate the radio earth from the mains safety earth, and this can be achieved when using either matching method shown above, albeit at the expense of some 'copper' loss in the windings. In Fig 3.12(a) a separate primary winding can be used instead of the auto-transformer method. When using the Fig 3.12(b) arrangement, a few turns of wire can be wound round the earthy end of the coil, but the number of turns must be determined by experimentation and this can be very fiddly to get right.

Practical inductors for 136kHz

Each of these inductors is variable with a maximum of 3 to 4mH which should be enough to match most low frequency antennas. If more inductance is needed, a fixed coil can be added in series with the variable one.

Tapped coil: A typical tapped inductor comprises 200 turns of 1.5mm (16SWG) plastic insulated wire wound on a 150m diameter former. The taps are at 100, 50, 25, 12, 6, 5, 4, 3, 2, and 1 turn, giving adjustment over a very wide range, whilst being capable of fine tuning.

The photograph opposite shows OK1FIG's loading coil which uses this type of design. He uses the lower turns for coupling to the transmitter, tapped for adjustment of the matching. Tuning and matching are achieved by selecting taps with 4mm sockets and banana plugs. It is important to make sure the transmitter is switched off when making adjustments. For a more permanent arrangement, the 4mm sockets can be replaced with soldered connections.

Wide range variometer: The construction method of a coil built by G3LNP is shown in **Fig 3.14**. The coil has an inductance range of 6:1 in three stages without gaps and is intended to be used with a separate toroidal matching transformer (see above).

Fig 3.14: G3LNP's variometer design allows a very wide range of inductance adjustable with no gaps

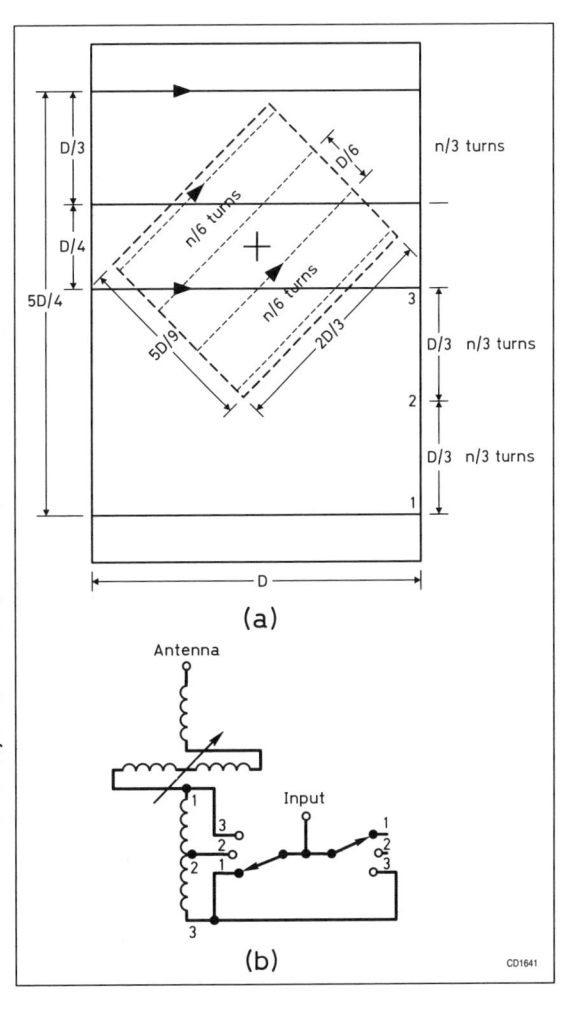

Fig 3.15: Detail of G3LNP's inner coil bearings. Note that this degree of engineering is not essential

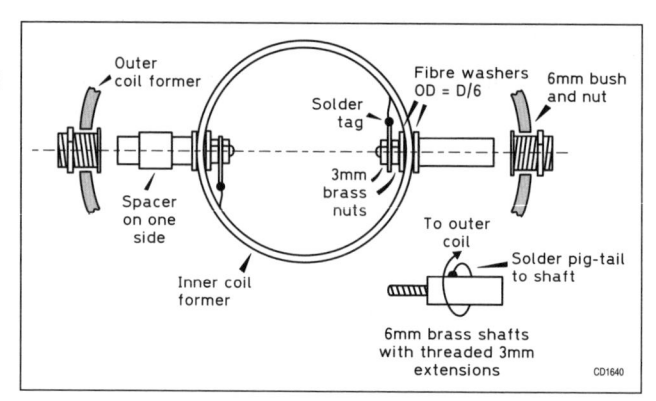

Taps 1, 2, and 3 provide respectively 2L/3 to L, L/3 to 2L/3 and L/5 to L/3 where L is the total inductance. Unused turns below tap 2 can be left open whilst unused turns below tap 3 should be short-circuited to minimise both voltage stress and losses.

The inductor typically has a total of 150 - 200 turns of 1.5mm diameter (16-17SWG) wire in a single layer. The number of turns chosen will depend on how large your antenna is, but it does no harm to have too many. The former is 200mm diameter and about 250mm long, but this is not critical - it depends largely on what is available.

Fig 3.15 shows the construction. The turns are divided between the outer main coil and the inner variometer coil. Each coil is split to allow space for the rotating mechanism. Construction is straightforward using 6mm shafts running in bushes salvaged from old volume controls and doubling as terminations for the flexible "pigtails" forming inner to outer coil connections. These mechanics are compatible with standard insulated shaft couplers and control knobs.

The important considerations are to make sure all the windings have the same direction and that rotation is restricted to 180 degrees. It is advisable to use a working tap which puts the inner coil in the additive inductance position to minimise losses.

Use the formula below to calculate the turns (n) on the larger coil for maximum inductance L (ignore the small coil):

$$n = \sqrt{50L/D}$$

where D is diameter of outer coil in inches and L is in µH. The distribution of the turns (in terms of n) is shown in Fig 3.15.

Losses in the inductor

The resistance of the loading coil results in the loss of a proportion of the transmitter power in the coil. The percentage loss of power is given by:

$$\% \text{Loss} = \frac{R_{coil}}{R_{loss} + R_{coil}} \times 100\%$$

or, in decibels, $\text{Loss(dB)} = 10 Log_{10} \left(\frac{R_{coil}}{R_{loss} + R_{coil}} \right)$

where R_{coil} is the coil resistance, and R_{loss} is the loss resistance of the antenna.

CHAPTER 3: ANTENNAS AND MATCHING

> **WARNING**
> Antenna tuning circuits carry very high voltages that can burn, as well as high currents that can cause injury or death. Always make sure that the transmitter is off - and cannot be keyed by anyone else - before touching tuner components, eg to change coil taps. Rapid changes in impedance may lead to the destruction of the power amplifier transistors.

Taking an inverted L antenna as an example, with R_{loss} of 40 ohms at 136kHz, and using a coil with a resistance of 15 ohms will result in 27% of the transmitter power being dissipated in the coil, while a coil with a resistance of only three ohms is used, only 6% of the power is dissipated in the coil. The loss in radiated power in decibels is 1.4dB for the first case and 0.25dB for the second. In either case, this loss amounts to only a fraction of an S- point at the receiving station, so the effect on overall system performance is minimal.

ON6ND shows off his well-engineered coil with its variometer and current meter

What is more significant is the power handling capability of the coil. Physically small coils are suitable for transmitter power levels of up to a few hundred watts. Larger coils with higher Q dissipate a smaller proportion of the transmitter power, and also have greater surface area for cooling.

Practical inductors for 472kHz

All of the information above can be used to design and build loading coils and variometers for use at 472kHz. In general, they will have lower inductance which means fewer turns and probably a smaller former. They also do not need to cope with such high power levels, especially voltage, so the engineering is much easier.

Tuning up

A Marconi antenna can be tuned up roughly by optimising the received signal. If possible, choose a constant ground-wave signal in or close to the band.

On the 136kHz band, DCF39 on 138.830kHz is useful for this, but it should be used only during daylight as sky-wave interference can produce slow fading.

Unfortunately, there are few big commercial signals close to 472kHz, but aeronautical beacons inside, or close to the band (see the chapter on Operating Practice) or broadcast stations at the lower end of the medium wave band can be a start. Alternatively, ask a fairly local station to transmit a beacon for you.

Then, using the largest taps first, vary the taps on the loading coil. When the strongest signal is received, alter the variometer position (or the fine tuning taps) to peak it. It may be necessary to make further adjustments to the taps so that the variometer tunes correctly. If the antenna fails completely to tune, make sure the earth is connected. If all else fails, tune your receiver to a beacon on a lower or higher frequency and see if resonance can be obtained there.

Once resonance is achieved within the amateur band, apply a little RF, perhaps from your driver, and measure the antenna current using one of the methods described in the measurements chapter.

Note that such tests should be carried out in a part of the band that avoids interference to other operators (if in doubt, ask) and that you should send a callsign at regular intervals. If you cannot avoid the more popular parts of the band, perhaps because your transmitter is crystal controlled, make sure you send callsigns often and listen periodically for any calls. Another option is to do this when most operators are at work.

Having re-peaked the antenna tuning, now is the time to tackle the matching. Turn the transmitter off, adjust the matching (coil taps, link winding size or matching transformer taps). Key the transmitter again and re-tune the antenna, noting whether the new matching setting is an improvement. Continue this until an optimum setting is found.

Gradually increase the amount of power applied to the antenna, monitoring the antenna current. If a sudden jump in current occurs (upwards or downwards), or an increase in power does not increase the antenna current, it is probable that something is breaking down, possibly flashing over. Carefully examine any visible parts of the antenna and its matching system and repair any problems. If flash-over is suspected, but no evidence can be seen, try testing the antenna after dark when any sparks will be more easily visible. Corona discharge is sometimes only very faintly visible, and may be easier to detect by the hissing sound it makes. Take great care not to damage your transmitter whilst doing this.

If all is well, you will have an amp or more (somewhat less on 472kHz) going into your antenna and a stable system that can be adjusted smoothly either side of its present tuning and matching.

Transmitting loops

An alternative to the Marconi vertical is a vertical loop of wire, often rectangular in shape with the lowest part a metre or so off the ground. **Fig 3.16** gives the basic arrangement. The loop has the advantage that it works well in locations where there are many vertical objects, such as trees, which would adversely affect a Marconi. It has the disadvantage that it is directional, with nulls of some 20dB at right angles to the plane of the loop.

Like the Marconi, the loop requires careful engineering in order to be efficient. The main difference is that Marconis have lots of volts whilst loops have lots of current - perhaps tens of amperes on the 136kHz band. As with any other small antenna, the efficiency is determined by the ratio of the radiation resistance to the loss resistance. It is vital that the loop is made as large as possible, and the loop resistance kept as low as possible.

Fig 3.16: Schematic of a typical LF loop antenna, showing a balanced capacitive potential divider feed

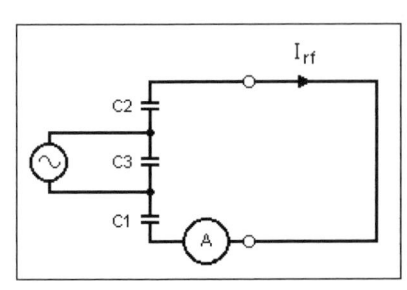

The radiation resistance depends on the loop area, not the shape, but some shapes require more wire (and hence resistive loss) than others. GW4ALG found a triangular shape convenient (**Fig 3.17**). A circular loop needs least wire and a square needs less than a rectangle. In practice, a rectangular loop is most often constructed as this will give the largest loop for the real estate and height available.

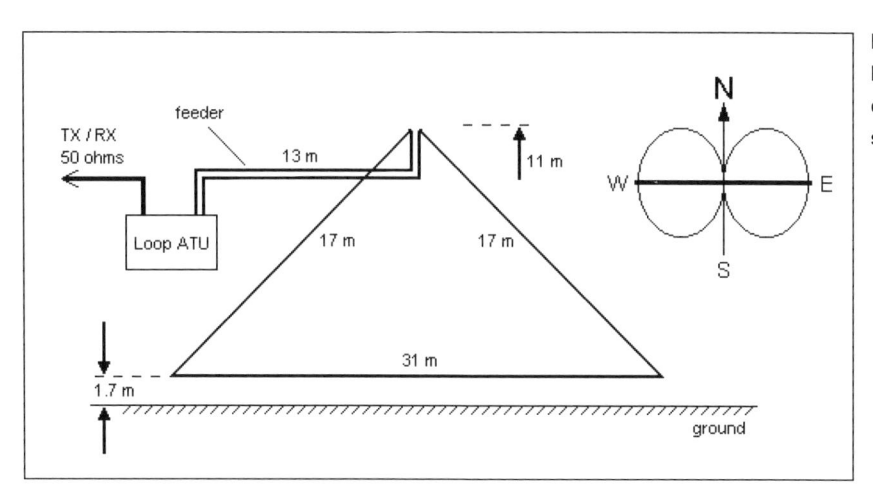

Fig 3.17: GW4ALG's loop antenna is a quite different shape

The calculations for evaluating your own loop can be found in the Measurement and Calculations chapter of this book. As a starting point a practical antenna used by G3YMC on 136kHz is described below.

At MF, a loop antenna should be competitive in efficiency with a vertical; however, at the time of writing, very few loop antennas have been tried on 500 or 472kHz.

The main advantage of transmitting loops is that the loop voltages are much lower than for the vertical (although they may still be in the kV region for high power stations), resulting in lower dielectric losses in objects around the antenna. This makes a loop a good choice for wooded surroundings, where many trees close to the antenna would lead to very poor efficiency with a vertical. This seems to be a common situation in North America, where several LF loop antennas have been constructed using branches of tall trees to support the loop element. Loops also do not rely on a low resistance ground connection, so may be an improvement where there is very dry or rocky soil. A disadvantage is that stronger antenna supports are required to support the thick loop conductor.

Feeding, tuning and matching

As a low frequency loop is very small compared with the wavelength, the polarisation is vertical and independent of the feedpoint, ie it can be fed at any convenient point. The loop can be considered to be an inductance which must be brought to resonance with a capacitor. The capacitors used for tuning, and perhaps matching too, are required to carry high currents and voltages, so must be of good quality to keep losses down.

Metallised polypropylene 'pulse' type capacitors have low losses in this frequency range, and are available in large capacitances and voltage ratings of 2500V or more. The limiting factor with loop tuning capacitors is often internal heating due to the RF current rather than the breakdown voltage. In large polypropylene capacitors, currents of a few amps are allowable, and loop voltages may reach a few kilovolts with transmitter powers typically employed at 136kHz, so series-parallel combinations of capacitors will often be needed. Components need not be so highly rated if the loop is used exclusively at 472kHz. More information on capacitors and their sources are in Appendix 2.

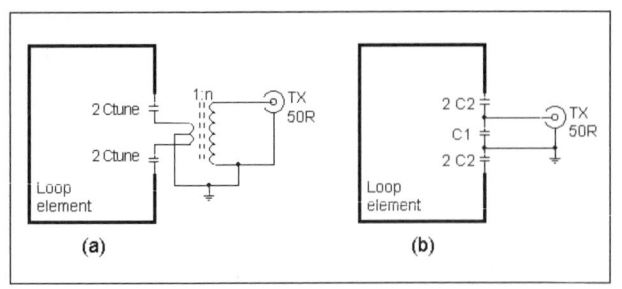

Fig 3.18: Methods of feeding a loop antenna: (a) Toroidal transformer similar to that used for feeding a Marconi, and (b) a capacitive potential divider

100 capacitors connected in series / parallel to match G3YXM's 136kHz loop [7]

Matching a loop antenna normally uses one of the circuits shown in **Fig 3.18**. **Fig 3.18(a)** uses a step-down transformer to match the loop R_{loss} to the 50 ohm transmitter output, and a series capacitance to resonate the loop inductance L_{ant}. L_{ant} in henries is given approximately by the formula:

$$L_{ant} = 2 \times 10^{-7} P \cdot Log_e \left(\frac{3440A}{dP} \right)$$

where P is the overall length of the loop perimeter (m), A is the loop area (m^2), and d is the conductor diameter (mm).

C_{tune} is therefore:

$$C_{tune} = \left(\frac{1}{2\pi f \sqrt{L_{ant}}} \right)^2$$

C_{tune} is often divided into two series capacitors as shown, to make the loop voltages approximately balanced with respect to ground. The required transformer turns ratio is $\sqrt{R_{load}/R_{loss}}$.

An alternative matching scheme uses a capacitive matching network, **Fig 3.18(b)**. The values of C1 and C2 are:

$$C_1 = \frac{\sqrt{\frac{R_{load} - R_{loss}}{R_{loss}}}}{2\pi f R_{load}}, \quad C_2 = \frac{1}{2\pi f \left(2\pi f L_{ant} - \sqrt{R_{loss}(R_{load} - R_{loss})} \right)}$$

Practical loop antennas

A detailed article on LF transmitting loop antennas can be found at [6], and descriptions of setting up practical transmitting loops are given at [7, 8].

GYMC's loop antenna

The loop used by G3YMC (**Fig 3.19**) was the result of site restrictions that made a Marconi difficult to implement [9]. He described the construction as follows:

The location is in a typical suburban estate with terraced houses on all sides and a back garden of 15m x 7m. It is not practical to use the front of the house so scope for [136kHz] antennas is limited. A loop antenna seemed the ideal answer. It was decided to use the whole of the garden for the antenna, which allowed a loop with a wire length of 45m, and an area of 100m². The first loop erected used ordinary 19/0.76 wire. However, the resistance of 45m of this is about 1.2 ohms, and it rapidly became apparent from the reports received (or lack of them) that the efficiency was very poor at about 0.001% with a radiated power of a few hundred microwatts.

The wire was then replaced with heavy duty loudspeaker cable with the pairs connected in parallel, to give an effective wire cross-section of 5mm² and a measured DC resistance of less than 0.1 ohms. In fact the RF resistance at 136kHz was rather higher at 0.65 ohms due to skin effect and other considerations.

The loop is a rather dog-legged construction. One end is supported by a 10m pole by the house, the other end is at only 5m, since rear guying of a pole there is not possible. The wire is run in a sort of rhombus between these points, with the lower sides running along the garden fence about 0.3m above ground. The loop is fed at one corner at ground level, via a matching box screwed to the wall of the house. The dimensions in **Fig 3.19** are given as a guide only.

The calculations on the loop revealed the following:

- loop area 100m²
- radiation resistance 13.5 micro-ohms
- effective loop series resistance 0.65 ohms
- efficiency 0.002%
- radiated power for 35W input 0.7mW
- Loop inductance 70µH

The loop is matched to the transmitter by the capacitive divider method shown in Fig 3.18(b). In this case the capacitor on the transmitter feed (C1) is 200nF and the tuning capacitor (C2) is 22nF. Polypropylene capacitors from the Philips 376 series were used. These are available in various voltage ratings up to 2kV. Surprisingly significant is that the change in temperature between early morning and mid day resulted in a resonance change of some 200Hz HF; in the cold winter mornings there was appreciable shift LF.

The bandwidth of the loop is quite small, and it is only possible to move the transmit frequency by some 100Hz either side of resonance before retuning is necessary. Tuning is carried out with several small switched capacitors inside the matching box selected with miniature toggle switches. If you put up a loop and find its bandwidth quite broad there is something wrong with it.

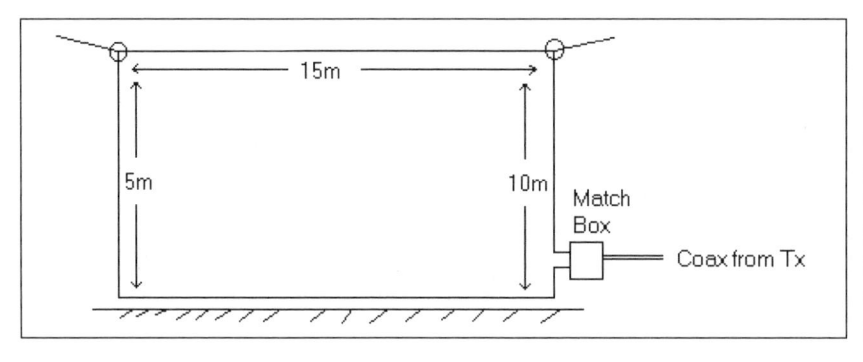

Fig 3.19: G3YMC's roughly rectangular loop antenna for 136kHz which helped him work five countries on CW with just 35W RF input

Measuring the current in the loop wire with a thermocouple ammeter was found to give a good guide to loop performance and enabled the actual value of the series resistance to be confirmed. Due to the magnitude of the current, which can be quite high, the test was performed at a reduced power level of 5W. At this power level a current of 2.6A was measured in the loop. The loss resistance can be calculated directly using Ohms Law and works out at 0.66 ohms (allowing for the DC resistance of the meter). It was encouraging that this agreed so closely with the computer simulations.

Note the magnitude of the current. At G3YMC's normal 35W it is nearly 8A. If the power were to be increased to 400W, a hefty 26A would flow. It is clear that there will be significant heating of the matching capacitors at this level and use of suitable capacitors is most important.

Comparisons between received signals on the loop and a 60ft longwire antenna showed stations in the line of the loop to be typically 10-15dB better on the loop than on the other antenna [at this site]. However stations in the null are some 10-15dB better on the longwire. This suggests the null is around 30dB.

The loop offers an antenna for small sites where large antennas are out of the question and the ground resistance is poor.

Earth antennas

Over the last few years there have been a few attempts to use the ground, or even the sea, as an antenna for low frequency amateur radio. None has been particularly successful, mainly because they have been compared (unfavourably) with the performance of large conventional antennas.

Recently, two things have changed that. One is the idea that an 'earth antenna' could be used in circumstances where it is impractical to erect a Marconi or loop, and a very inefficient antenna is better than nothing. The other concerns experiments at VLF (see later chapter) where practical antennas are often extremely hard to achieve.

Experiments with an earth electrode pair

G3XBM has carried out experiments on the VHF, LF and MF bands and has been surprised by the results. The following is taken from a much longer article [10]:

Like many, my home is not blessed with a large garden. Small Marconi verticals and wire loop designs have been tried [on LF/MF] with varying degrees of success. In the past, I have used earth-electrode pairs for my VLF work, so I thought it was worth trying these on the new 472kHz band to see how such a system would perform.

I was not expecting very good results but was amazed how well it worked, both on transmit and receive.

An 'earth electrode pair' consists of a pair of earth rods, in my case about 1m long copper earth stakes, driven into the soil about 20m apart. The output of the transceiver or transverter is connected directly to the two earth rods using ordinary PVC covered wire. I used 32 x 0.2mm (1mm^2) wire.

The wires are simply laid out across the grass, as shown in **Fig 3.20**. It is important that the space between the two electrodes is not bridged by other earthed structures: the best results will be when they are far apart and 'in the clear', far from any other pipes, buried metalwork or cables.

In my case, the loss resistance of the earth electrode system works out as being 66 ohms. This is virtually the same as the feed impedance as the radiation resistance is around 0.017 ohms. Therefore I am able to directly connect my nominally 50-ohm rig into the earth-electrode antenna. Depending on the impedance of any a particular earth electrode pair, some form of matching may be necessary. The current was measured with an in-line antenna current meter [the one in this book].

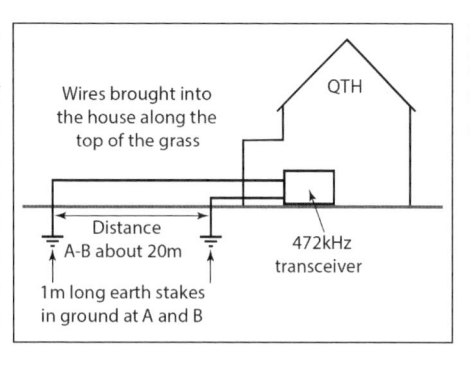

Fig 3.20: G3XBM's basic earth electrode antenna setup in his garden

To test the performance of this earth electrode pair, it was necessary to do some comparisons with other antennas. I used a small 9m high, top-loaded Marconi vertical tuned to 472kHz using a fixed inductance near the top of the vertical section and additional inductance in series at the bottom to bring the whole antenna to resonance.

To compare the field strength from each antenna I ran a series of WSPR beacon transmissions over a period of many hours with each antenna. I made a record of the received signal to noise ratio at many reporting stations. To avoid any chance of one antenna disturbing the other, only one antenna was erected at a time. The RF output from my transmitter, was around 10W although the radiated power is a fraction of this.

Based on several days of observations of my signals by stations up to 701km distant, I established that the earth electrode antenna is between 2 and 14dB down on the Marconi antenna, with worst results from stations at right angles to the structure. For stations 'in line', the difference in performance was only a few dB. In other words, this antenna is not as good as my Marconi vertical, but it still works pretty well and gets reports from a long way away (see **Table 3.1**).

It is equally effective on receive, with good signals received from many stations across Western Europe on WSPR and on CW.

So how does it work? Although it is not possible to be certain, I believe the best explanation is that the earth-electrode pair antenna acts as a large vertical loop in the ground: current flows from one earth rod (A) into the ground and returns to the other rod (B) via a series of diffuse paths within the soil and rock beneath the structure (**Fig 3.21**). How far down and out the signal spreads will depend on soil chemistry and on the geology of the rock beneath. In my case the ground is a light alkaline soil over chalk bedrock that is less than 2m below the surface.

It is possible to improve the performance of this structure by raising the

Reporter	SNR (dB)	km	Azimuth
DL-SWL	-28	701	83°
PI4THT	-30	448	88°
PA0A	-25	417	73°
PA3ABK	-30	306	98°
M0LMH	-30	223	329°
G0KTN	-27	210	242°
G4AGE	-25	151	316°
G0MQW	-23	123	225°
M1GEO	-28	79	184°
G3ZJO	-20	79	270°
M0BMU	-30	69	210°
G0BPU	-23	67	110°
G7NKS	-16	46	240°
G8KNN	-11	12	248°
G4HJW	+7	9	180°

Table 3.1: WSPR spots received when using <10mW EIRP from the earth-electrode antenna

Fig 3.21: How the current flows in the ground to form a virtual vertical loop.

height of the connecting wires, forming part of the loop in the air as well as in the ground. This is indeed how this antenna started out, until someone suggested I try it with the wires laying on the ground in the grass: performance was almost unchanged and it is much easier to string a piece of wire across a lawn than elevate it up in the air. The small difference in performance as a result of elevating the 'above ground' part of the loop adds credence to the 'large loop in the ground' theory.

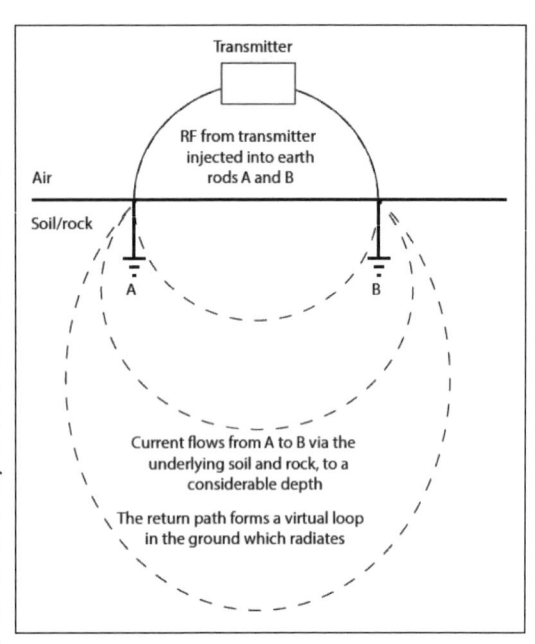

Clearly, most people will want to radiate as much power as possible, up to the legal limit. However, many are, like me, unable to erect large antennas and are prepared to accept a reduced EIRP, especially if the 'antenna' system becomes ridiculously simple. With the performance of the earth-electrode averaging just 8dB below that of my 9m high Marconi, the performance is certainly good enough to use it for semi-local CW QSOs out to at least 50-75km, even though my EIRP is only around 5mW from the 10W transmitter.

From my earlier work at 136kHz, I know this structure is able to radiate a signal over a considerable distance, even with very low EIRP levels. At 136kHz, if one is prepared to accept the compromises, such an approach is a useful alternative when compared with a very large 'in the air' antenna with large, low loss, loading coils that may need to be in the 4-6mH range.

The earth-electrode pair can be a very useful alternative 'antenna' for the 472 - 479kHz band. Although results will very much depend on local soil, rocks and the degree of metal clutter in the garden, it is certainly an antenna to try when larger antennas are not possible.

Why use 50Ω on LF/MF?

Unlike amateur HF and VHF/UHF equipment, low frequency transmitters and antennas may not have impedances of exactly 50 ohms. It is, however, convenient to use this impedance for cabling, and to transform up or down as appropriate. This makes matching easier, allows the use of conventional test gear and reduces cable losses which can be significant if mis-matched, even at low frequencies.

Antenna supports

In striving to achieve the required height for an efficient low frequency antenna, all manner of supports have been used, including an apartment block, lighthouses and a church tower.

A more practical alternative, is a tall tree. Trees make useful supports, cause no problems with the neighbours and do not require planning permission. However, they can absorb quite a lot of the RF, especially in summer. It is interesting to note that LF antennas near trees show a marked change in resonance during spring and autumn. They also tend to move when it is windy, so some method must be used to avoid the antenna or the tree branch breaking. For instance, bungee cables, used for securing items on car roofs, can be placed in series with the antenna wire and/or the rope that goes over the tree branch.

A catapult or bow and arrow can be used to propel an object over a high branch, attached to a monofilament line. This can be used to haul up a much thicker rope attached to the antenna. Alternatively a pole can be erected using the tree trunk as a support.

In the absence of enough tall buildings or trees, one or more masts must be erected. Note that this may require planning permission. In general, the mast should be at the maximum practical height. It may be possible to use a mast that can be raised and lowered when required, either telescopically or folded over.

Should the mast be insulated from the ground? Well, ideally it should, and it could then be used as part of the antenna. This is often used by commercial LF stations, but it is quite difficult to engineer, bearing in mind the large voltages present on short verticals. A more practical solution is to earth the mast and ensure that any antenna wires are at least a few metres away from it. The worst option is to have a mast that has resistance to earth, say via brackets fixed to a wall or through a concrete base, as this will dissipate RF in a way that varies from day to day. If you are in doubt, earth it.

Care needs to be taken with guy wires. If they are insulated from ground, capacitive coupling to the antenna element can result in high RF voltages being present on the guy wire, presenting a hazard for anyone coming into contact with them (this can also apply to other nearby conducting objects, such as ladders, garden furniture etc). This can be avoided by grounding them. However, if the antenna wire (for instance the vertical element of a Marconi) runs inside the guys, grounding them may result in a partial screen around the antenna generally reducing the effective height. Insulating rope used instead of wire for guys is one possible solution.

One method of using a tree as a support is to strap a pole to the trunk

Masts made of an insulating material may be a useful option. Telescopic fibreglass masts can be purchased [11] and they are suitable for portable use, though rather too flexible for a permanent installation. Roach poles used by fishermen may also be employed in an LF antenna.

Temporary antenna supports

A helium filled balloon can make a useful temporary support for a vertical wire. Several 400mm balloons, of the type used for children's parties, can be used in a bunch. This is an option for anyone unable to erect a permanent high mast, though obviously it is only available on a windless day. The lower part of the antenna can be supported conventionally on a pole which allows

G3LDO's fold over 16m mast

the balloons to carry less wire, reduces the possibility of the wire tangling in nearby objects and keeps the high voltages away from the ground. GW4ALG used this method to make many DX CW contacts on 136kHz from his tiny garden.

Portable operation at LF is more difficult than from a fixed station, because it can take a long time to raise the antenna to a decent height. Kite supported antennas have been used with great success [12, 13], although some skill in flying a kite must be acquired before using it to support an antenna. The wire can either be trailed beneath the kite, or more usefully can be made part of, or replace, the supporting 'string'. Given the right kite and good wind conditions, a 60m vertical wire (the maximum kite height allowed in the UK) can be supported in a stable way for hours. Safety is an important issue and no part of the kite should be able to touch an overhead wire. Furthermore, when transmitting, the wire will carry harmful voltages and must not be anywhere near members of the public. There are additional height regulations when operating close to an airfield.

References

[1] ON7YD's antenna pages. *http://www.strobbe.eu/on7yd/136ant/*
[2] 'Radiation from an antenna', P Dodd, G3LDO. *http://pe2bz.philpem.me.uk/Comm/-%20Antenna/-%20AntennaDesign/Radiation/Basic_EM.htm*
[3] *Radio Communication Handbook*, RSGB.
[4] *http://rsgb.org/main/technical/emc/emc-publications-and-leaflets/*
[5] Loading coil spreadsheet: *http://www.wireless.org.uk/tap_coil.xls*
[6] *www.we0h.us/Amateur_Radio_stuff/Transmitting-Loops/Bill-Ashlock-documents/TX-Loop-Antennas-for-the-1W-Lowfer-Band-Part-2B.doc* - article on LF transmitting loops
[7] Loop antenna at GM3YXM: *http://www.wireless.org.uk/loopy.htm*
[8] *www.w1tag.com/XESANT.htm* - transmitting loop used by WD2XES
[9] G3YMC's loop page. *http://www.davesergeant.com/loops.htm*
[10] 'A novel 472 - 479kHz antenna', Roger Lapthorn, G3XBM, *RadCom*, March 2013, RSGB
[11] Moonraker, Unit 12, Cranfield Road Units, Cranfield Road, Woburn Sands, Bucks MK17 8UR. Tel: 01908 281705. *http://www.moonrakerukltd.com/*
[12] G3XDV's portable operation using a kite. *http://homepage.ntlworld.com/mike.dennison/index/lf/gw3xdv/october99.htm*
[13] G3YXM' portable operation using a kite. *http://www.wireless.org.uk/136gm.htm*

4

Receive antennas

In this chapter:

- Why use a receive antenna?
- Using a transmitting antenna
- Antennas for receive only
- Positioning a receive antenna
- Noise on feeders
- Practical loop antennas
- Active whip antennas
- Reducing electrical noise

WHY USE AN ANTENNA DESIGNED only for receiving? Firstly, because a receive-only antenna does not require the complexity and sheer scale of one used for transmitting, some people will want simply to receive on LF and not transmit at all. There is a great deal of pleasure to be had in optimising a low frequency receiving station and then giving DX stations reports or contributing to the experimental data being compiled on propagation. Any of the antennas described here are suitable for serious receiving. It is easier than ever for the operator of a receive-only LF/MF station to provide reception reports via the Internet. This can simply be by email, by using the DXCluster system, or by using the automatic web reporting systems incorporated on some modes such as WSPR and Opera. More on web reporting can be found in the chapter on operating practice.

Secondly, an antenna designed to be efficient for transmitting may pick up unwanted signals and noise that can be reduced in amplitude by using an antenna optimised for receiving, possibly in conjunction with interference-reduction tools described later in this chapter.

The newcomer to receiving on low frequencies will find that strong signals can be received on a random untuned long wire, giving the impression that this type of antenna is adequate. However, this type of antenna is prone to picking up local electrical interference and strong out of band signals. Much better performance can be obtained from a properly matched receive antenna.

Using a transmitting antenna

Those equipped for transmitting will find that their antenna will work well as a general purpose receive antenna. It may, however, pick up local electrical noise which is particularly prevalent on these frequencies, and will certainly have no controllable directivity. So it is often useful to supplement the main station antenna by having one solely for receiving.

LF TODAY

It is often the case that local noise originates outside the amateur's home, propagating along mains or telecomms cables, via which it is coupled to the antenna. Re-positioning the antenna can reduce the noise level, but this is not often an option with a large transmitting-type antenna.

The best approach under these conditions is to locate a separate, compact, receiving antenna where local noise pick-up is minimised. With suitable design, quite small antennas can provide a signal-to-noise ratio limited only by the external band noise, so there is no compromise in receiving sensitivity.

Antennas for receive-only

Dedicated receiving antennas for the LF and MF ranges can be classified as whips and loops (**Fig 4.1**). A whip is just a small vertical antenna, and responds primarily to the electric field (E-field) component of a radio wave. A loop develops a signal at its terminals primarily due to induction by the magnetic field component (H-field) component of the radio wave. At a particular location, the fields generated by local noise sources may be chiefly E- or H-field, so using the appropriate type of receiving antenna can give a substantial reduction in noise. However, the nature of RF fields is such that both E- and H- fields are always present to some extent even if one dominates the other, so neither type of antenna is a 'miracle cure' for all noise problems. Nevertheless, experimentation with the type and position of the receive antenna can often yield major reception improvements.

Fig 4.1: Some low frequency receiving antennas: (a) loop, and (b) active whip

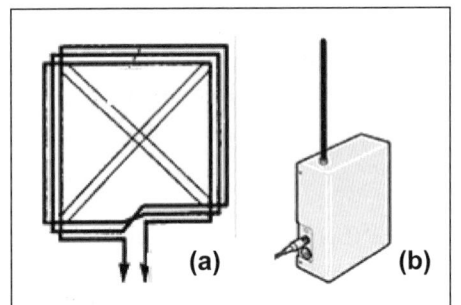

A whip antenna is omnidirectional. A loop has a figure-of-eight directional pattern, with two deep nulls at right angles to the plane of the loop, and maximum reception along the plane of the loop. This can be a disadvantage if the loop is fixed and the wanted station is in the null, but small loops can be rotated by hand or using a rotator suitable for a VHF antenna.

On the other hand, the directional characteristic can be extremely useful in nulling out interference from man-made sources, either local (for instance switch-mode power supplies which can be a problem on both 136kHz and 472kHz) or distant (for instance Loran or the strong utility stations on 136kHz), provided there is a considerable difference in direction between wanted signal and unwanted noise.

The E-field produced at a location by a received signal varies considerably due to the local screening effects of buildings, trees, etc. near the receiving antenna, so the signal level at the output of a whip antenna is strongly affected by its location. The H-field is much less affected, so loops can operate well when tucked away in a confined location. This also makes loops a better choice when repeatable signal level measurements are wanted, for example when making field strength measurements.

A whip antenna can actually be a self-supporting whip, or a short wire element. It may be tuned in the same way as a transmitting antenna, or, for example, using the preselector circuit shown in the Receivers chapter.

Active whip antennas

Alternatively, active whip antennas (sometimes called 'E-field probes') are untuned, but use a high-impedance buffer amplifier to match the whip element to the low-impedance receiver input.

Active whips give wide-band reception, and can be very small - often the whip element is around 1m long. Due to their wide-band nature, the buffer amplifier must be well designed to avoid intermodulation, especially if located close to broadcast stations. A popular active whip design by PA0RDT is described later in this chapter. The signal level at the output of an active whip is quite sensitive to its location; when the whip is at ground level, output will be much lower than when elevated to a position where it is not screened by surrounding buildings, trees, etc. This loss in signal level can be compensated for by increasing the size of the whip element. Being essentially vertical antennas, both tuned and active whips require a good ground connection.

Large and small loops

Several varieties of loop antennas exist, including tuned loops, active loops and also 'terminated' loops. For many years, tuned loop 'frame' antennas have been used by medium wave DXers, and these can also be effective for LF/MF amateur use.

A typical design for 472kHz uses a square wooden frame with about 1 metre sides, wound with around 15 turns and tuned with the paralleled sections of a 500pF + 500pF variable capacitor. A tap at one or two turns from the grounded side of the winding matches the loop to a low-impedance receiver input. For 136kHz reception, about 30 turns can be used instead. The signal output from a loop like this is a small fraction of a microvolt, and a preamplifier (such as those shown in this chapter and the one on Receivers) will be needed for all but the most sensitive radios.

Some amateurs have built much larger loops; the signal output is roughly proportional to the area of the loop, so a larger loop with an area of several square metres can eliminate the need for a preamplifier. However, it is more difficult to position a large loop to minimise noise pick-up.

The tuned loop has a high Q, and is quite effective in filtering out unwanted signals. The drawback of this is that the bandwidth of the loop is very narrow and it will frequently require re-tuning, which is somewhat awkward when the antenna is some distance from the shack. It is possible to make the loop tunable from the receiver end of the feeder; see the 'Lazy Loop' described later.

Fig 4.3: Example of a well-constructed LF loop by PA0SE

Fig 4.2: The K9AY EWE antenna has an omnidirectional response

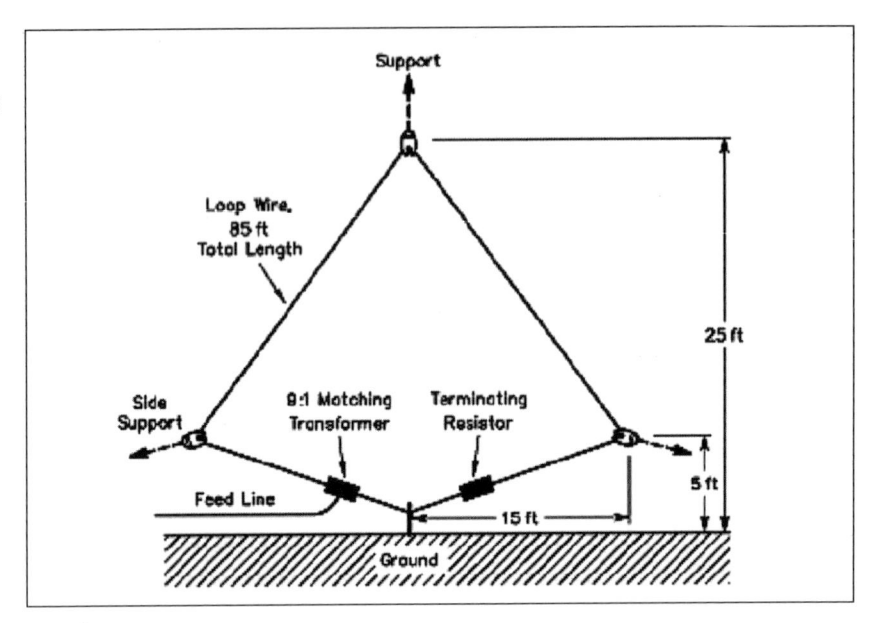

Un-tuned active loops can be made to have a very uniform response over a wide bandwidth; they are widely used professionally for measuring field strength for this reason. Active loops suitable for the LF/MF range are commercially available, for example the LFL1010 antenna from Wellbrook Communications [1].

Several types of un-tuned 'terminated loop' antennas have been developed, the best known being the K9AY (**Fig 4.2**) [2] and 'EWE' antennas [3]. These use a relatively large wire element simultaneously as a loop and a vertical antenna, with a terminating impedance that combines both signal components at the receiver input to give a unidirectional (cardioid) pattern. These antennas are capable of excellent results at 136kHz and 472kHz. They do however require a fairly large area of ground in a noise-free location for good results.

Positioning a receive antenna

The most important thing in achieving effective noise reduction using a dedicated receiving antenna is to find the optimum position for it.

Whip antennas usually give best signal-to-noise ratio when clear of surrounding obstructions and as high up as possible, preferably on a mast of some sort. This will also keep the antenna as far as possible from mains noise sources, which tend to generate greatest noise levels inside and near buildings. Whips are vulnerable to picking up noise from other resonant antennas; if a whip is close to the main LF/MF transmitting antenna, the main antenna loading coil should be open-circuited while receiving, for instance using a relay at the 'cold' end.

Loop antennas are particularly prone to picking up the magnetic fields generated by noise on mains cables so should also be kept as far from mains wiring as possible. However, the noise level will often change greatly over distances of only a few metres, so moving the loop around within the available area will often locate a 'sweet spot' where noise levels are low. The directional property of the loop may often be used to null a local noise source; the apparent direction of the null on the local noise may also change over a short distance, so in some cases it will be desirable to have two or more loops in different positions to receive signals from different directions.

As can be seen from these remarks, all types of receiving antenna perform best outdoors; indoor receiving antennas almost always give disappointing results in this frequency range. If an indoor antenna is unavoidable, it will again be worthwhile experimenting with different locations inside the building, or even within the same room, since positions further from mains wiring and noisy appliances will usually have significantly less noise.

Noise on feeders

A problem that may affect any type of antenna used for reception is noise currents flowing along the outer braid of coaxial feeders. Since the receiver end of the cable is normally connected to mains earth, the feeder provides an alternative path for mains noise currents to flow through, particularly if there is an additional RF ground at the antenna end (a 'ground loop'). Noise currents flowing along the feeder will in turn induce noise in the receiving antenna. The best way to check for this problem is to monitor the noise level with a battery-operated receiver, making sure that no connection exists between the mains ground and the antenna and receiver.

Loop antennas do not depend on an earth connection for operation, so keeping the loop and feeder insulated from ground will minimise the amount of noise current present on the feeder. In at least one case, a high noise level in a loop antenna was caused by long grass coming into contact with the loop and creating a conductive path for noise currents.

Vertical antennas, including receiving whips, require a ground connection, preferably as close as possible to the antenna. The level of noise current in the feeder can be greatly reduced by inserting a 1:1 ratio RF isolating transformer into the feeder, breaking the connection between mains and RF ground, while still allowing signals to pass along the feeder normally. The PA0RDT Mini-Whip design includes such an isolating transformer in the DC power feed unit at the receiver end of the feeder. A similar result can be achieved with a transmitting antenna by using a matching transformer with separate primary and secondary windings to isolate the feeder at the antenna end.

Practical loop antennas

A traditional tuned loop antenna is shown in **Fig 4.3**. It consists of several turns of wire wound onto a wooden cross-shaped frame, typically about 1m2, parallel tuned by a variable capacitor (usually a combination of fixed and variable to make tuning less fiddly).

The receiver input is fed via a low impedance single turn winding. The output of the loop is small, and a low-noise preamplifier will normally be required, such as the one shown in **Fig 4.4**.

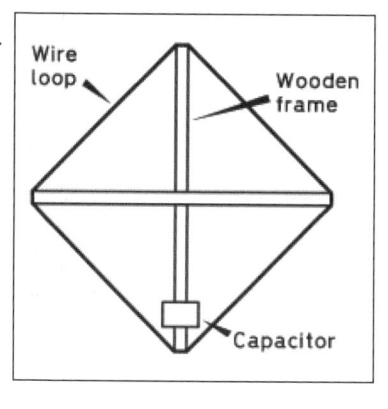

Fig 4.3: LF tuned loop antenna. For 136kHz, use 30 turns and a 1000pF capacitor; for 472kHz use 10 turns and 500pF

Some thoughts on loop antennas
Abridged from a post from Jim Moritz, M0BMU, on the RSGB LF Group

For small VLF - MF receive loops, the available signal power from the loop depends mostly on the area of the loop. You get maximum power delivered to the receiver when the loaded Q with the receiver connected is half the unloaded Q. This happens when the impedance connected to the loop has a resistive component equal to the loss resistance, and a capacitive reactance equal to the inductive reactance of the loop, ie resonance and conjugate match.

To a first approximation, the number of turns does not affect the signal power level, it only affects the design of the impedance matching. With very small area loops you want to try to achieve conjugate matching in order to maximise sensitivity. But with relatively large area loops (of the order of one square metre or more) and a suitable low-noise receiver or preamp you can get enough signal so that band noise is the limiting factor even with quite a large mismatch. In particular, mismatching to reduce the Q reduces bandwidth, which is often very useful.

At LF/MF it is usually possible to make preamps whose noise level is either below the band noise, or below the thermal noise level generated by the loss resistance of the loop, so the loop itself is normally the limiting factor. With several square metres area, you can get ample signal level without bothering to tune to resonance, and have really wideband antennas. The EMF induced in a given loop is proportional to frequency, so the output decreases at low frequencies, but since the band noise level increases at low frequencies, this usually isn't a problem.

How many turns the loop has is largely a matter of convenience for the tuning and impedance-matching point of view. The traditional high-Q loop has many turns so that it can be tuned to resonance with available variable capacitors, but there are several disadvantages, such as being easily de-tuned, susceptible to capacitive noise pick-up, requires remote tuning and mechanically complicated.

Small single-turn loops have the drawback of requiring a large tuning capacitor. For instance, my 136kHz one metre square "bandpass loop" design [see this chapter] has a 0.4uF tuning capacitance. But since such a narrow band is to be covered, the tuning can be fixed, and the loaded Q is quite low so this isn't really a problem, and the loop is a simple square of tubing that is not significantly detuned even if laid on the ground.

Some people have made quite large, high-Q tuned loops - the main justification for this would seem to be to get a large signal output so that a receiver with poor sensitivity can be used directly without a preamp. The high Q helps to prevent the receiver being overloaded by out-of-band signals. The same result could be achieved with a small/wideband loop and a suitable preamp/preselector.

A large-area wire loop can be connected directly to a low impedance receiver input with reasonable results as a wideband antenna - or a considerable improvement in matching can be obtained using a wideband transformer. The loop inductance can be incorporated as part of the filter/matching network, as I have done in the "bandpass loop" designs. Low-pass designs are also quite feasible. I am currently using 4 x 5m single-turn wire loops incorporating low-pass filters matched to 50 ohms, with cut-off frequencies around 550kHz to reduce local broadcast signals. These loops give plenty of signal over the 10 - 550kHz range without retuning.

Small loops have the advantage that they are easier to re-position away from noise sources. The height above ground, or the presence of trees and buildings has little effect on the signal level.

Whether the loop is big or small, narrow- or wide-band, it has the same figure-of-eight pattern, so the nulls can be used to reject local and distant QRM. Unless your real estate is big enough for LF phased arrays, the loop is the only feasible LF directional antenna. At M0BMU, MF reception would currently be almost impossible without loops due to wideband QRM radiated by nearby broadcast stations.

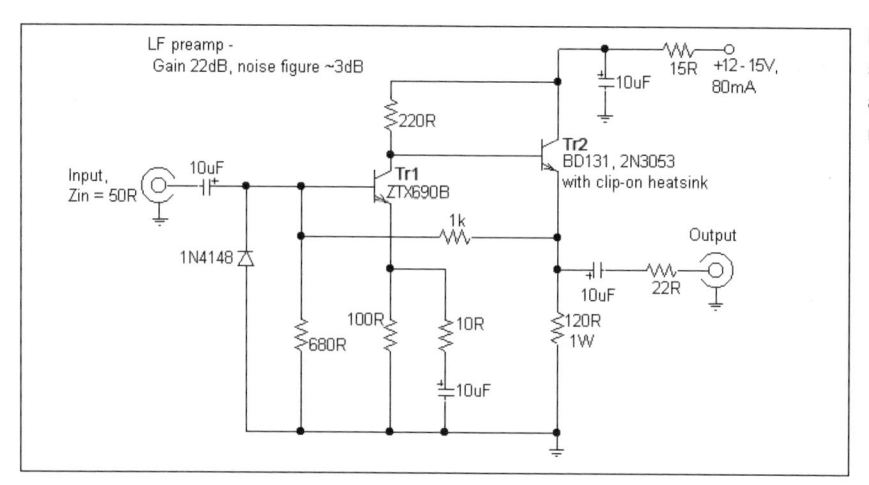

Fig 4.4: Preamp with 50-ohm input suitable for loop antennas

The Q of the loop is typically 100 or more, so re-tuning will be required even within the narrow low frequency bands. This selectivity is very useful in reducing intermodulation due to strong out-of-band signals. A number of amateurs have used much larger tuned loops for reception, which achieve higher output signal levels and so can dispense with the preamplifier, at the expense of being more bulky.

A loop preamplfier
This untuned preamp in Fig 4.4 has an input impedance of 50 ohms, and with a 50 ohm load gives 22dB gain. The noise figure is roughly 3dB, which is low enough for most things, and has quite good strong signal performance - it can cope with 100mV RMS input without clipping, so the receiver will almost always be overloaded before the preamp. An alternative pre-amplifier can be found in the Receivers chapter.

The output impedance is more or less that of the 22 ohm resistor, which ensures stability with reactive loads, so it is not fussy about what load it is connected to. The bandwidth with the transistors shown is about 5MHz, but it is really meant for the LF/VLF range. TR1 can also be a ZTX650 with similar results, also a 2N4401 or 2N2222 are OK but with 1 or 2dB more noise. A BD135, BFY52 or 2N3053 work well for TR2 - this runs quite warm so a clip-on heatsink is desirable. Extra filtering of the 12V will be required if there is significant noise on the supply.

M0BMU 'Lazy Loop'
An alternative to the conventional multi-turn loop is the 'lazy loop' of **Fig 4.5**. This uses a large single-turn loop, the area of which is around 10 - 20m^2. The shape is not at all important, and it can be normal insulated wire slung from bushes or fence posts, etc, hence the name! The loop can be fed through a coax feeder, allowing it to be positioned remote from the shack to reduce noise.

The other great advantage of this system is that the critical tuning components are nice and dry in the shack, no remote control, no waterproofing of the amplifier, just twiddle and go! It also gives a useful amount of front end selectivity. This tuning arrangement is not optimum from the point of view of minimising

LF TODAY

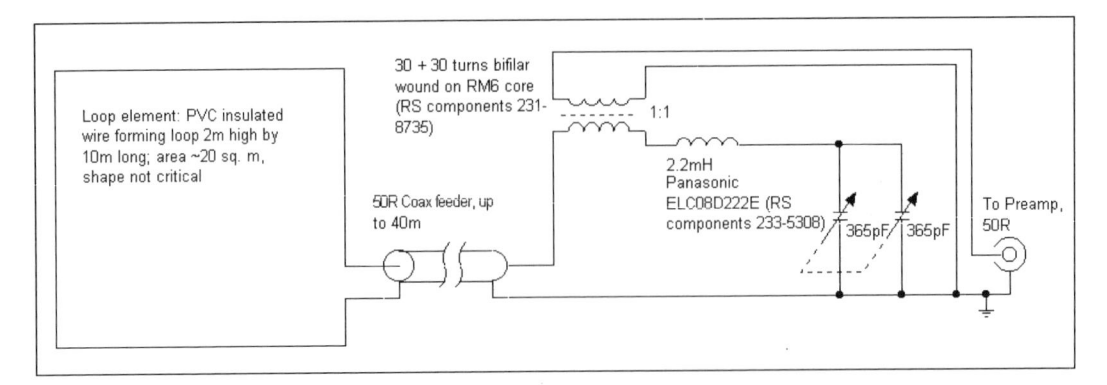

Fig 4.5 'Lazy loop' with its tuning arrangement. The 2.2mH inductance is intended for the 136kHz band, but the loop will work at 472kHz with a smaller value

[Diagram: Loop element: PVC insulated wire forming loop 2m high by 10m long; area ~20 sq. m, shape not critical. 50R Coax feeder, up to 40m. 30 + 30 turns bifilar wound on RM6 core (RS components 231-8735), 1:1. 2.2mH Panasonic ELC08D222E (RS components 233-5308). 365pF / 365pF. To Preamp, 50R]

losses, but due to the large loop area the signal-to-noise ratio is more than adequate. It is also possible to use somewhat smaller, multi-turn loops.

Again, a low-noise preamp, such as the one in Fig 4.4 will be required. The preamp has been used with several different types of antenna, but the arrangement in Fig 4.5 is quite useful. It is a fairly large single turn loop, which is series tuned by an inductor and variable capacitor. The inductor should be reasonably high Q - I used two 1mH ferrite cored chokes (RS components 233-5291), with a Q of about 80. The transformer allows one side of the tuning capacitor to be grounded - almost any high-permeability ferrite core should work OK. The tuning part is located in the shack, and connected to the loop with coax; this can be quite long, since it effectively just increases the inductance of the loop, which in any case is much smaller than the 2mH tuning inductance.

The area of the loop required depends on how 'deaf' the receiver is. If sensitivity is inadequate, a bigger loop can be put up. The area is the important factor which decides the signal level; height and shape are not critical. This means it only takes a few minutes to put up, and can use any available supports - the original loop was just slung between a fence post and the branches of a bush, with no additional insulation. A wire thrown over a small tree also worked fine - hence the 'Lazy' title. It could also be useful as a receiving antenna for portable operation.

Of course, the antenna should be mounted as far as possible from noise sources, such as mains wiring. If one end is made moveable, the loop can be turned to null out QRM. The tuning inductance can easily be changed to cover different frequency ranges, including 472kHz.

M0BMU's bandpass loop for 136kHz and 472kHz

The frequent need to re-tune a high Q loop is something of a drawback. The Q can be reduced by adding resistive loading, but unfortunately this also reduces the signal level available to the receiver. An alternative approach is to combine the tuned circuit formed by the loop with other capacitors and inductors to form a bandpass filter. This results in a wider bandwidth, while at the same time improving rejection of out-of-band signals.

Another drawback of conventional loops is the multi-turn winding, which is hard to weatherproof. A single-turn loop made of tubing is more convenient and robust, and has quite high Q, even though very large tuning capacitance is needed. The following design is based on a 1m x 1m loop element made of 15mm

CHAPTER 4: RECEIVE ANTENNAS

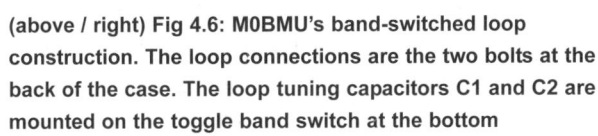

(above / right) Fig 4.6: M0BMU's band-switched loop construction. The loop connections are the two bolts at the back of the case. The loop tuning capacitors C1 and C2 are mounted on the toggle band switch at the bottom

copper water pipe (as shown in **Fig 4.6**) and using the bandpass filter principle to achieve coverage of the full band without re-tuning.

Fig 4.7 shows how similar circuits, with the same inductance values, are used for both 136kHz and 472kHz bands, The transformer in the 472kHz circuit is not critical; any transformer giving a low loss and 200:50 ohm transformation at 472kHz could be used. The -3dB bandwidths of the prototypes were approximately 10kHz for the 136kHz version, and 37kHz for the 472kHz version.

S1 simply switches the single loop element between the two circuits. The band-change switch selecting the loop resonating capacitor must have low contact resistance in order not to increase the losses in the loop, so a 4-pole, double throw toggle switch with three poles connected in parallel was used to select the loop capacitor.

Table 10.5 shows details of the components. The 600µH adjustable inductors are the primary windings of 'Toko' or similar 455kHz IF transformers; the types

Fig 4.7: M0BMU's band-switched LF/MF loop

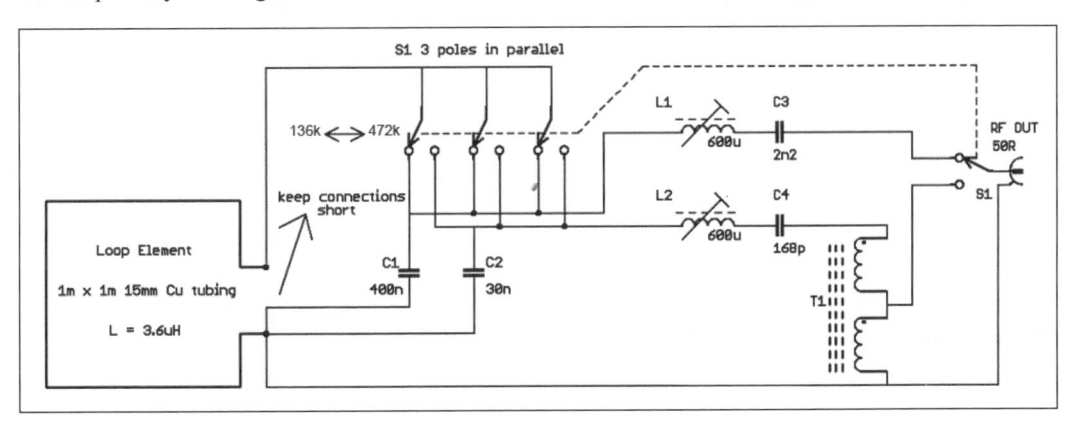

LF TODAY

Table 4.1: Component notes for bandpass loops

Capacitors		Stable, low-loss types:
C1		4 x 100nF, 100V metallised polypropylene in parallel (capacitors selected during alignment)
C2		2 x 15nF polystyrene in parallel (capacitors selected during alignment)
C3		2.2nF polypropylene
C4		150pF polystyrene + 18pF in parallel
L1, L2		Primary winding of 455kHz 10mm 'Toko' IF transformer with 180pF tuning capacitor or similar; capacitor removed.
S1		4-pole double-throw miniature toggle switch
T1		18 bifilar turns on FT-50-43 (5943000301) 12.7mm dia, = 850 toroid, or similar

that have 180pF capacitors have a suitable inductance value that is adjustable over a fairly wide range. The internal ceramic capacitor must be disconnected or removed; in the Toko types, this is most easily done by carefully breaking up the ceramic capacitor in the moulded plastic base with a pointed implement. The primary winding is usually connected to the two end pins of the row of three pins on the IFT base - check with an ohmmeter to find the largest winding resistance. Other coils with similar inductance and a Q >50 could be used.

The loop element is made from 4 x 1m lengths of 15mm copper water pipe, joined in a square with 90 degree solder elbows. One side of the loop is cut in the middle, the cut ends flattened and drilled, and brass bolts passed through and soldered into place. The bolts pass through the wall of a plastic box containing the tuning components, and connections are made using solder tags. This gives a good low-resistance connection. The connections to the loop tuning capacitors should be as short as possible and direct to the loop terminals, especially for the band-switched version - see Fig 4.6. The loop element is attached to a wooden support using plastic pipe clips.

Alignment consists of adjusting the resonant frequencies of the loop and auxiliary tuned circuits. This is best done during assembly by temporarily configuring the tuned circuits as parallel or series traps as in **Fig 4.8**; when the resonant frequency is equal to the source frequency, a sharp dip in detector level will be seen. First, the parallel resonant frequency of the loop itself is adjusted to be

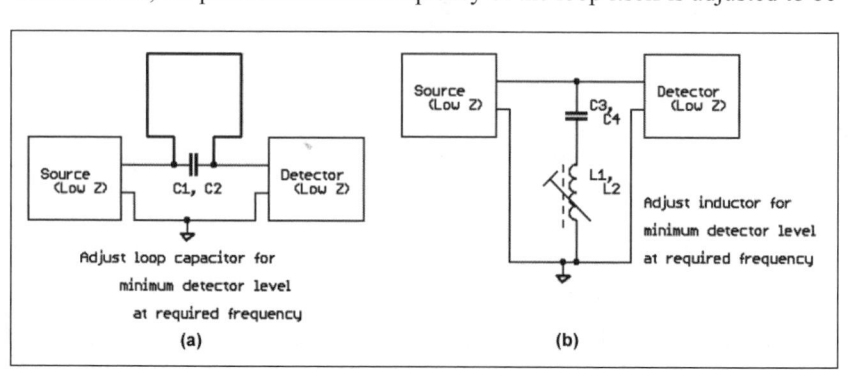

Fig 4.8: Adjusting the resonant frequency of (a) loop, (b) auxiliary tuned circuit

close to the centre of the band of interest by selecting a suitable parallel combination of capacitors (**Fig 4.8(a)**). The auxiliary tuned circuit is then adjusted to an identical series resonant frequency by itself (**Fig 4.8(b)**). Then the connections between the tuned circuits and the output are made to complete the circuit; no further adjustment should be required. Nominal resonant frequencies for the two bands are 137kHz and 475kHz, although deviations of a couple of percent from these values are not serious due to the fairly wide passband - the main thing is that both tuned circuits are set to the same frequency. Note that, when tuning the Toko inductors, these tiny ferrite cores can be saturated by quite low signal levels; about -20dBm from the source is safe.

The output signal level is quite small, and a low-noise pre-amplifier such as the circuit of Fig 4.4 is used. Sensitivity is then more than adequate to hear the band noise levels. The loops are not very sensitive to de-tuning; even a large metal object like a step-ladder has little effect unless it is nearly touching the loop.

These designs were originally published as a longer article which can be found at [4].

Active whip antennas

Quite short wire or whip antennas provide adequate signal to noise ratio at LF when matched to the receiver input by a high impedance buffer amplifier. The response of the resulting Active Whip antenna, or E-field antenna, is broadband, and can extend from the VLF range to the VHF range, depending on the amplifier used. The preamplifier is located at the base of the antenna, and its DC supply is usually fed up the coax.

The whip element is often around 1 - 2m long, and the small size makes the antenna easy to site in an electrically quiet location. As with the loop antennas, it is important to experiment with the antenna location to find a site that has a low noise level. Since all signals over a wide frequency range are presented to the buffer amplifier input, good dynamic range is important, and overload problems may occur if there are high-power broadcast stations nearby.

Unlike loop receiving antennas, the signal output level from an active whip depends strongly on its position, partly due to the screening effect of surrounding buildings and other objects, and partly on the height of the whip element above the ground plane. Greater output will be obtained if the whip is mounted on a mast, or on the roof of a building, between 3 and 5m above ground is ideal.

The PA0RDT Mini-Whip antenna

The PA0RDT-Mini-Whip© was designed by Roelof Bakker, PA0RDT, and has been built and used successfully for LF and MF reception by numerous amateurs. Good receive performance extends from 10kHz to over 20MHz. The compact size of this antenna also makes it very suitable for portable operation.

The Mini-Whip 'whip' element is in fact a small piece of copper-clad board. The shape is not important provided capacitance to ground is similar, and other whip element construction can also be used, eg a small metal box with the pre-amp inside. Tests were performed to optimise the size of the whip element, and the design achieves good sensitivity while maintaining maximum overall output at about -20dBm to prevent receiver overload. The buffer amplifier is optimised for good strong-signal performance. Second order output intercept has been

LF TODAY

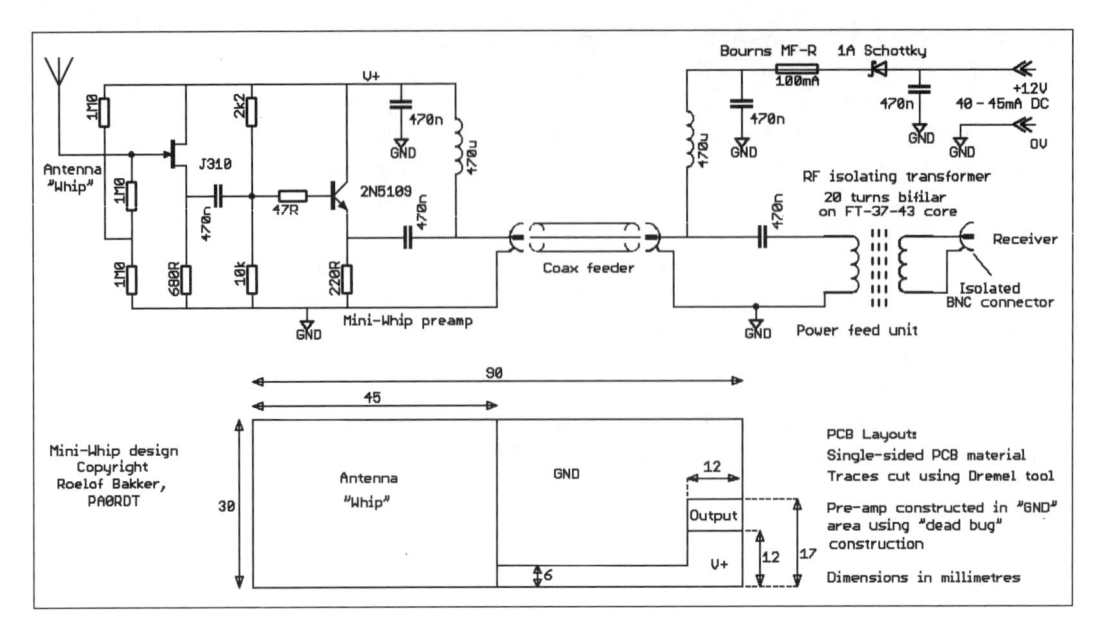

Fig. 4.9 The PA0RDT-Mini-Whip©

measured by AA7U as being greater than +70dBm, and third order intercept greater than +30dBm.

Power is fed from a 12 - 15V DC supply to the Mini-Whip via the power feed unit and the coaxial feed line, which can be up to 100m long. The power feed unit includes an RF isolating transformer, which reduces noise due to ground loops, but this is not always essential.

The buffer amplifier is constructed 'dead bug' style on the ground plane half of the board next to the whip element, which is formed by the other half of the board (see **Fig 4.9**). The complete circuit board is mounted inside a 100mm long section of 40mm plastic drain pipe, with two end-caps. One end-cap carries an insulated BNC connector to which the circuit board is soldered.

PA0RDT notes that the electric field from most interference sources is largely confined within the building. For best results therefore, the Mini-Whip should be mounted on a non-conducting pole in a position well clear of buildings. Grounding the outer braid of the coaxial feeder to a ground rod installed close to the point where the feeder enters the shack also helps to reduce interference generated by noise currents flowing on the feeder.

A more detailed article can be found at [5]. A ready-made Mini-Whip can also be purchased from PA0RDT [6].

Ferrite rod antennas

A compact, directional antenna can be made using a ferrite rod (or rods) and several operators have used them on the 136kHz and 472kHz bands. Efficiency on the lower frequency is very dependent on the type of rod and some experimentation is needed. Output is low so an amplifier is almost always required.

M0BMU experimented with a ferrite rod antenna which had more than adequate sensitivity, but he found the narrow <1kHz bandwidth to be a problem. It was necessary to re-tune when changing frequencies within the band, also the ferrite used caused considerable centre frequency drift with changes in ambient

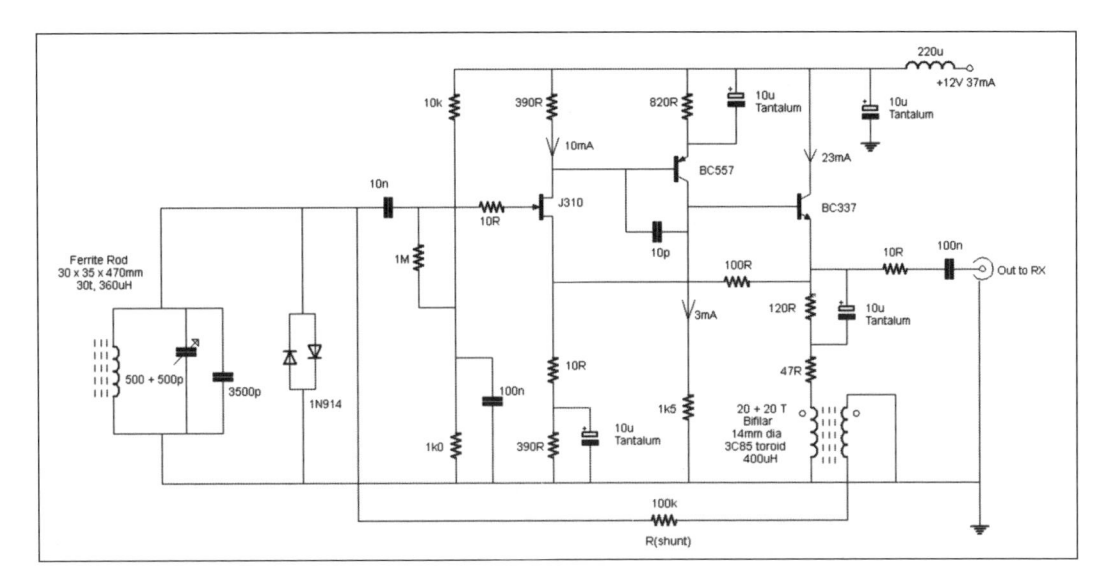

Fig 4.10: M0BMU's amplifier for a ferrite rod antenna

temperature. Since it is usually desirable to have the antenna some distance from the receiver, and remote tuning is a nuisance, a wider bandwidth of about 5kHz would be useful.

The bandwidth could be increased by adding a damping resistor to reduce the rod antenna Q, but unfortunately this also reduces the signal-noise ratio. Approximately, the signal voltage is proportional to the Q, but the thermal noise voltage produced by the loss resistance is proportional to the square root of the Q, so the signal level decreases more quickly than the noise level as the Q is reduced. For the rod on question, the desired bandwidth increase would result in about 7dB increase in noise floor, which would be marginal. What was needed was a loading resistor with reduced thermal noise.

DF6NM suggested one way to do that is to use a preamplifier with a feedback network that defines its input resistance to provide the loading. The preamp in **Fig 4.10** achieves this by using a shunt feedback resistor around an amplifier with a well-defined inverting voltage gain to provide the load resistance. The equivalent input resistance $R_{in} = R_{(shunt)} / (1-A)$; in this case gain A is -10 and $R_{(shunt)}$ is 100kohms, giving R_{in} of about 9k. However, the noise voltage at the input caused by $R_{(shunt)}$ is also reduced by about $1/(1-A)$, and is about 10dB less than if a 9k resistor was connected directly across the input.

Compared to the original ferrite rod antenna circuit, the result was that the bandwidth was increased to about 5kHz, and a preset tuning adjustment was adequate. The output noise level was still well below the external band noise, even under quiet band conditions. The basic idea is quite adaptable to different gain, impedance levels, frequency and so on.

Reducing electrical noise

The old adage "If you can't hear 'em, you can't work 'em" is as true at LF as on any other band. The limiting factor is often the background noise level which divides into atmospheric noise and local noise. This is dealt with in some detail in the chapter on operating, together with how to locate noise sources.

Fig 4.11: Noise cancelling array

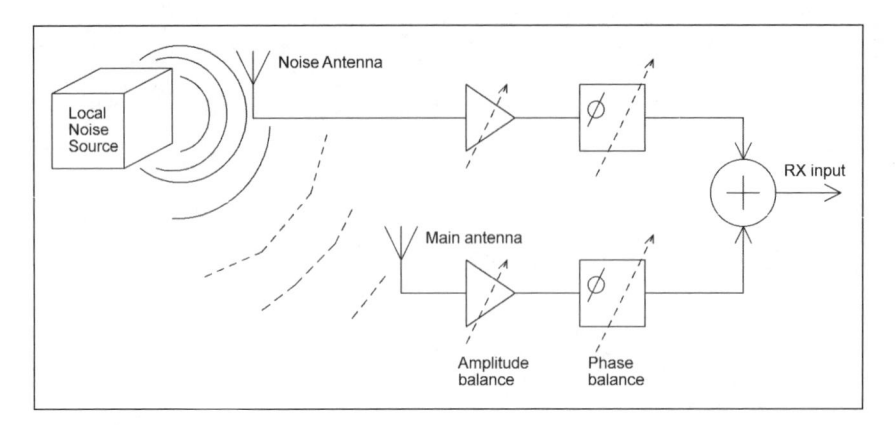

An obvious way to reduce the effects of local noise is to locate the antenna as far as possible from the noise sources, and this is good practice in any case.

Noise or persistant interfering signals may be reduced by introducing a 'null' in the direction of the interference. In fact, this is the only way to deal with distant noise sources such as static or strong commercial stations. One way to achieve a null is to use a directional receiving antenna, such as a loop as described above.

For local noise sources, noise cancelling may be effective. This relies on being able to position two antennas so that one is relatively much closer to the noise source as in **Fig 4.11**. Distant signals will received by both antennas at a similar level, but the 'noise' antenna will have a relatively higher level of the local noise present. Suitably attenuating and adjusting the phase of the noise antenna signal before summing the outputs of both antennas results in cancellation of the local noise, with relatively little change to distant signals. To be successful, this scheme requires that the noise originates predominantly from a single source; it is unlikely to be practically possible to arrange that multiple noise sources will have the correct amplitude and phase to all be cancelled simultaneously.

Cancellation can be achieved with a very simple circuit such as one by G3GRO in [7], but these can be tricky to adjust. The canceller can be made more versatile by providing it with adjustable gain, and phase shift that can be varied over a full 360 degree range. Buffer amplifiers can be used to isolate the gain- and phase-adjusting networks, making these adjustments independent of one another. A noise canceller of this type has been in use at M0BMU with a variety of receiving antennas at 136kHz and 472kHz. The circuit used is shown in **Fig4.12**.

The circuit is based on five identical high input impedance unity gain buffer circuits (**Fig 4.13**). Coarse adjustment of gain between -12dB and +12dB is provided at the two antenna input channels via tapping points on step-up/step down auto-transformers before being applied to the input buffers. Resistive loading of the transformers ensures a fixed input impedance close to 50 ohms independent of gain.

The input buffers each drive RC variable phase shift networks. The dual-gang 'phase balance' pot is wired differentially so that as the phase shift in one channel increases, the other channel decreases. This gives an overall phase adjustment of about +/-120 degrees. A further switched 0/180 degree phase

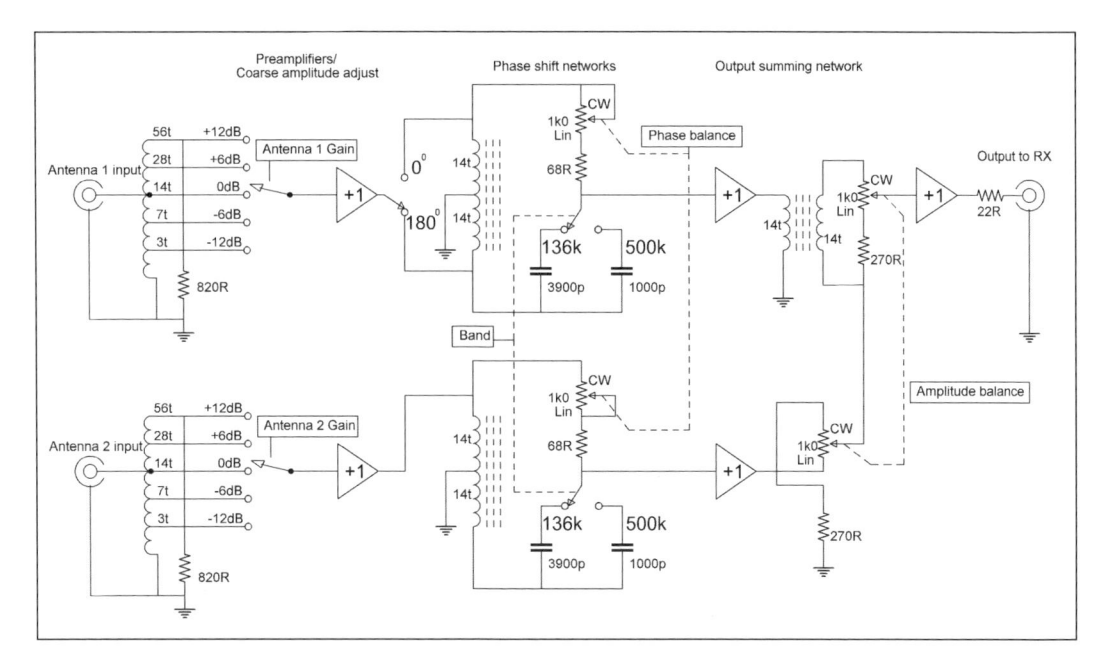

CHAPTER 4: RECEIVE ANTENNAS

Fig 4.12: M0BMU noise canceller

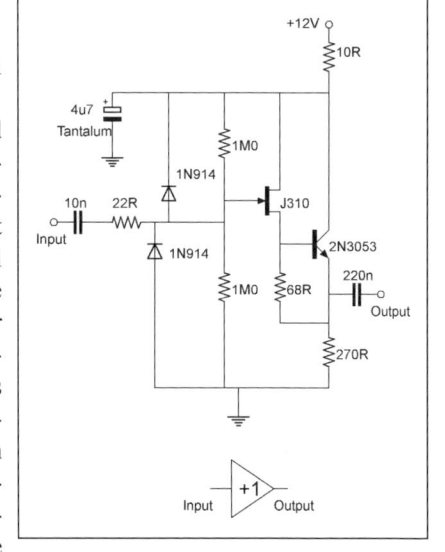

shift is provided by inverting one channel, so that a full 360 degree range is covered with overlap.

Output from the phase shift networks is buffered and applied to a gain adjustment network. The 'amplitude balance' pot provides approximately -10dB to 0dB gain variation in each channel, and again is wired differentially so that as the pot is rotated, the signal amplitude in one channel increases while the other decreases. Together with the coarse input gain adjustment, unwanted signals differing by over 30dB at the antenna inputs can be adjusted to a null, permitting a wide range of receiving antennas to be used. Signals in the two channels are summed by connecting the gain-adjusted outputs in series, one being made floating with respect to RF ground by an isolating transformer. The combined output passes through a final buffer to a low-impedance receiver input. The transformers shown in Fig 4.12 are all wound on high permeability ferrite toroids with A_L of approximately 4000nH/turn, similar to RS components part number 232-9561.

Fig 4.13: High impedance buffer used in M0BMU noise canceller

The noise canceller has been used for suppressing local noise sources, and to produce directional nulls on distant signals, mostly using the bandpass loop antennas described above, together with an un-tuned vertical antenna. Active whip and terminated loop type receiving antennas also work well. High Q resonant antennas are not well suited to noise cancelling schemes, since the signal phase and amplitude, and hence the depth of the null, is very sensitive to any alteration of the resonant frequency. Also, the bandwidth of the null is extremely small, and the controls must be re-adjusted for even a slight change in receiver frequency.

The system can be set up as follows. For nulling distant interference using the loop/vertical combination, the loop is first oriented for maximum received interference level (plane of the loop directed towards the source), or, if two sources are to be nulled, on a bearing mid-way between the interference sources. For cancelling local noise sources, one antenna is positioned to maximise noise pick-up, and the other for minimum noise. The amplitude controls of the noise canceller are then adjusted with each antenna individually connected in turn to obtain as close as possible to equal unwanted noise levels at the receiver from each antenna. Then both antennas are connected to the canceller, and the phase balance control adjusted for a null in the noise level. A few iterations of adjusting the amplitude and phase balance control should then result in a deep null.

References

[1]   Active loop antenna: *http://www.wellbrook.uk.com*

[2]   *http://www.hard-core-dx.com/nordicdx/antenna/loop/k9ay/* - K9AY loop antenna

[3]   *http://www.dxzone.com/catalog/Antennas/Receiving/EWE/* - about EWE loop antennas

[4]   *www.wireless.org.uk/BPloops2.pdf* - Bandpass receiving loop antennas for LF and MF.

[5]   *http://dl1dbc.net/SAQ/Mwhip/Article_pa0rdt-Mini-Whip_English.pdf* - the PA0RDT miniwhip antenna

[6]   E-mail: *roelof@ndb.demon.nl*

[7]   *www.carc.org.uk/pdf/Archive_003.pdf* - G3GRO 136kHz noise canceller.

5

Generating a signal

In this chapter:

- Signal-frequency VFOs
- Crystals and ceramic resonators
- Dividing a variable oscillator
- HF transmitter as a signal source
- Direct digital synthesis
- GPS locking
- Ultimate2 QRSS kit

ANY LOW FREQUENCY transmitter must start with a signal source. Because the allocations at 136kHz and 472kHz are so narrow, CW operation could be carried out with a crystal controlled transmitter, but being restricted to a single frequency is a serious practical limitation. Most operators use some form of variable oscillator, especially if they intend using modes that have nominated frequencies as centres of activity.

Some time should be spent considering the type of oscillator to be used as it depends on a number of requirements. In particular:

- Frequency stability
- Frequency accuracy
- Modes to be used

A CW-only station will need relatively low short-term frequency stability and low setting accuracy. On the other hand, QRSS120 must stay within 0.05Hz for hours at a time and its frequency must be known to a high degree of precision. DFCW will require a means to shift the frequency in a single step, whilst WSPR needs continuous control of the frequency. Other modes may have different requirements such varying the amplitude or phase of the signal source.

Even when buying a commercial low frequency transmitter - there are a few, see the Transmitters chapter - it is important to check whether its oscillator meets the specification for the mode(s) you intend to use.

More on the requirements of various popular low frequency operating modes can be found in the Modes chapter later in this book.

Signal-frequency VFOs
VFOs operating at the transmitter output frequency are not popular on these bands where most operators want the option to use modes requiring higher stability.

LF TODAY

Fig 5.2: Part of a transmitter designed by PA0SE. The 4060 chip divides the crystal to the 136kHz band

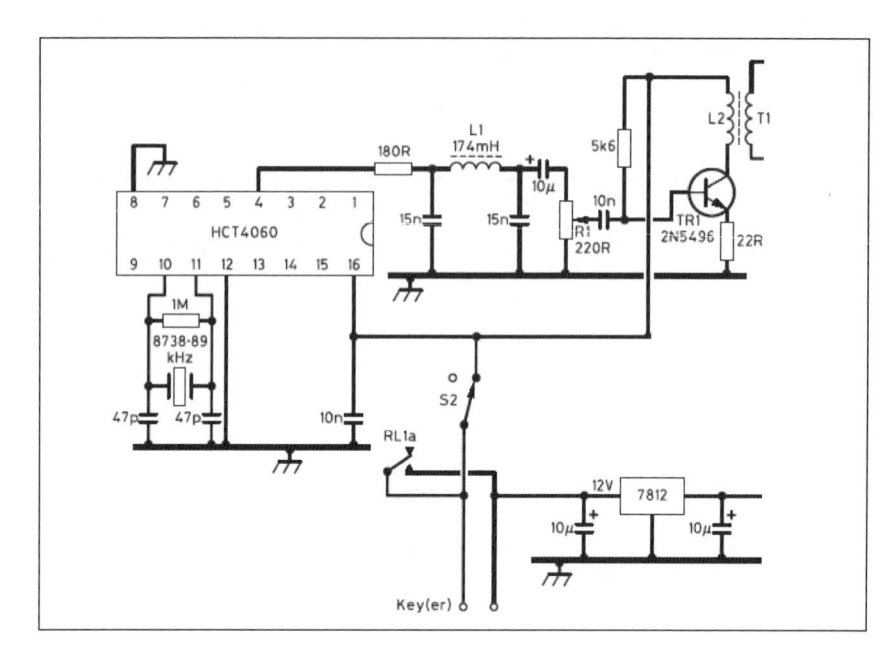

Crystals and ceramic resonators

The simplest way to generate a high-stability signal is by using a crystal oscillator, though signal frequency crystals for either the 136kHz or 472kHz bands are likely to be expensive and difficult to obtain.

Fig 5.3: 137 kilohertz ceramic resonator VFO

An HF crystal oscillator can be employed to provide an oscillator of very high stability and accuracy by using a digital divider chip. An example is shown in **Fig 5.2**, where the 8738.89kHz crystal frequency is divided by 64 by the 74HCT4060 IC to obtain output at 136.545kHz. The output frequency

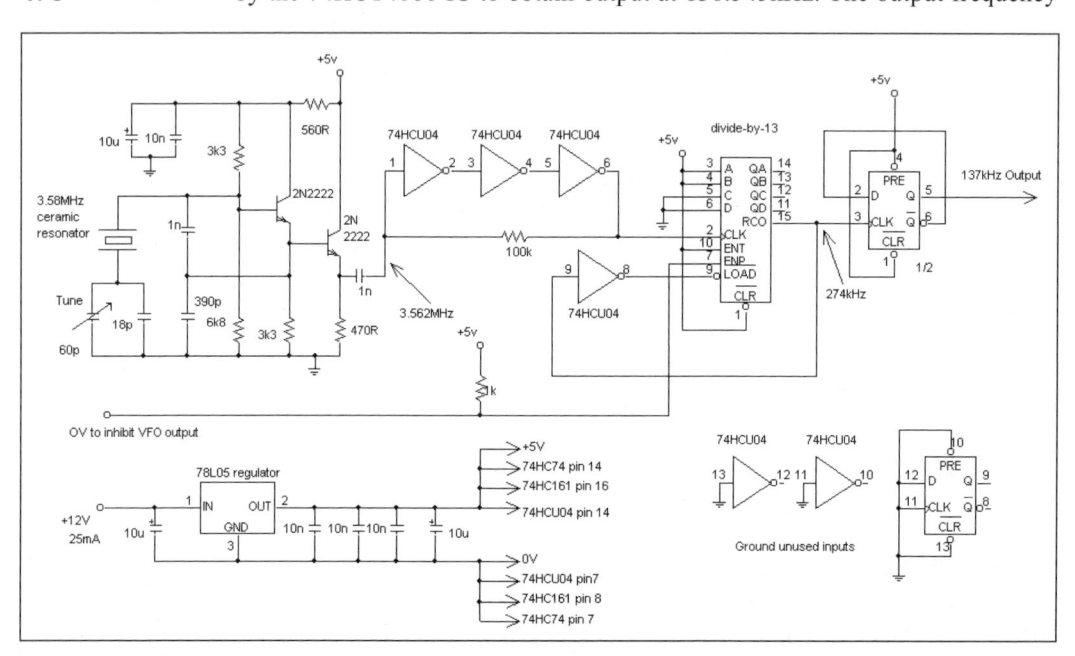

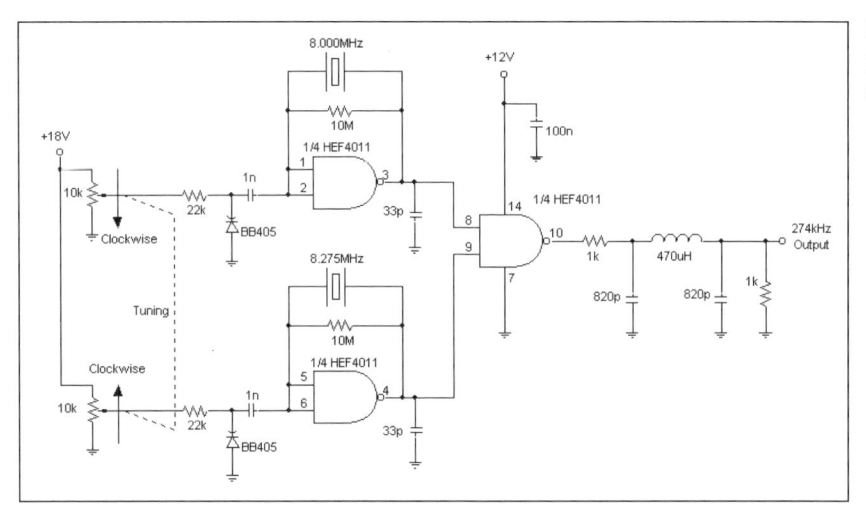

Fig 5.4: 'Crystal mixer' VFO from G0MRF transmitter

is fixed, but the stability is excellent, since any drift at the HF crystal frequency is also divided by 64. Provided the correct frequency crystal can be obtained, this oscillator is suitable for QRSS modes with dot lengths of up to 120 seconds.

GW3UEP's 1W transmitter uses a crystal oscillator employing a cheap 7.6MHz crystal to divide by 16 to 475kHz. It can be found in the Transmitters chapter.

An extension of this method, but with reduced stability is to divide a ceramic resonator oscillator to provide a variable oscillator with coverage of several kilohertz. **Fig 5.3** shows a version for the 136kHz band.

A more useful variable frequency signal source uses two HF crystal oscillators roughly 136kHz or 472kHz apart. These can be old crystals, in the 4MHz to 15MHz region, salvaged from your junk box - perhaps from an old VHF transceiver, found at a rally, or purchased new for a few pounds. One crystal oscillator is pulled in frequency by a series capacitor (sometimes an inductor is included, but this may reduce stability) to achieve a VXO with a range of a couple of kilohertz - a much greater amount than achievable using a low frequency crystal. The two crystal oscillators are mixed and filtered to produce an output covering a useful chunk of the band.

Fig 5.4 shows an example of this technique used by G0MRF in his class-D project which is featured in the Transmitters chapter. The same principle can of course be used to generate a 472kHz signal.

This approach is suitable for relatively low stability modes such as CW, Opera and QRSS3.

Dividing a variable oscillator

An HF VFO can be fed into a divider to produce a stable and variable source. This technique is used by G3YXM in his 136kHz transmitter; the circuit diagram includes the oscillator and can be found in the Transmitters chapter.

For his simple 472kHz 25W and 100W transmitters (see the Transmitters chapter), GW3UEP designed a VFO that divides by eight from 3.8MHz (**Fig 5.5**). It has a low component count yet will produce a stable enough signal for CW, Opera and

LF TODAY

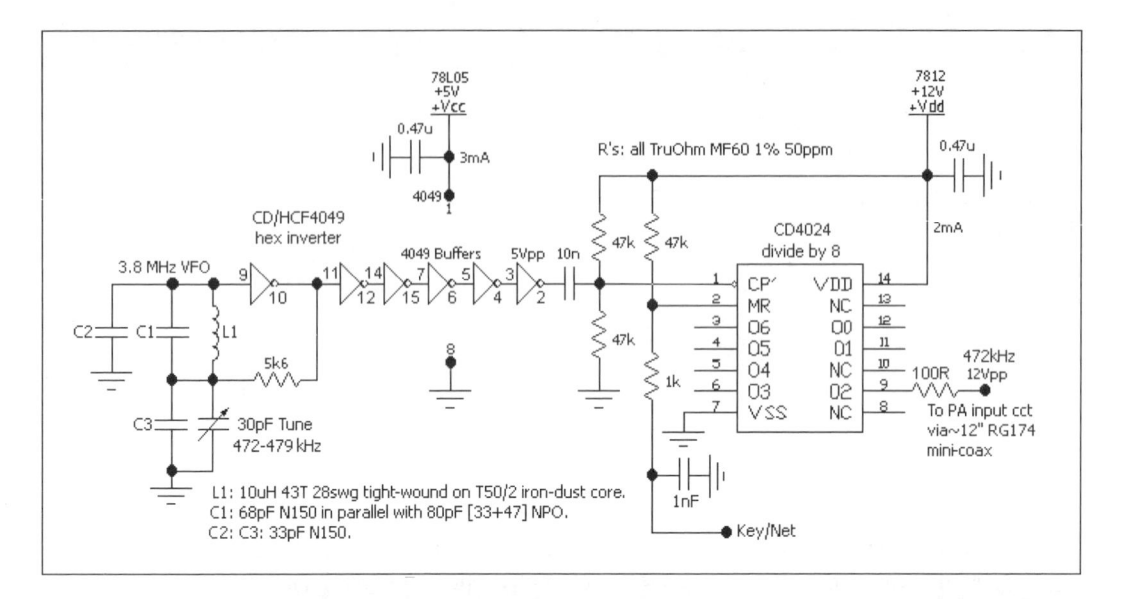

Fig 5.5: GW3UEP's stable VFO for the 472kHz band matches his 25W and 100W transmitters (see later chapter) The photograph show construction details

QRSS. A prototype is shown opposite. The oscillator runs continuously. Simple CMOS keying is achieved using the divider 4024 reset-line. Whilst using the companion 100W transmitter the keyed RF envelope is free from spikes and glitches, minimising key-clicks. It is important to take care to ensure maximum stability of the 3.8MHz oscillator. Inductor L1 is laquered then hot glued in place 4mm above the ground plane and clear of metalwork. Note the temperature coefficients specified for the capacitors in Fig 5.5. Non-ferrous (eg brass) screws/fixings should be used in the vicinity of the toroid. The VFO box is separate from the PA unit in order to avoid thermal coupling and temperature change. More details can be found at [1].

CHAPTER 5: GENERATING A SIGNAL

Dividing an HF transmitter
The transmitter in your existing HF station can be used as a VFO, by using a digital frequency divider (see above) to obtain low frequency output. This can give excellent stability and full band coverage. It may also be possible to generate narrow-band FSK using the transmitter in SSB mode.

It is convenient to choose a frequency division ratio of 10:1 or 100:1, so that the rig frequency display provides the correct readout (but with the decimal point moved). Most current HF transceivers inhibit transmission outside the amateur bands, but in many cases simple modifications extend the tuning range to convenient frequencies such as 13.600MHz or 4.720MHz (modification details for many rigs are available from the manufacturers and user web sites). The HF rig power output should be kept low, and connected to a suitable dummy load and attenuator to terminate the transmitter output and provide the correct level to drive the divider. Some transceivers have a 'transverter output' that can be used.

An HF transmitter as a signal source

The methods described previously are suitable for modes involving on-off keying, such as CW or Opera. With some modification they can be used with modes requiring a single frequency shift, such as DFCW and RTTY.

By using an SSB transmitter as a frequency source makes it easy to transmit a variety of data modes, for instance WSPR. Note that some modes need the LF/MF power amplifier stages to be linear. See the Modes chapter for the transmitter requirements of the modes commonly used on the low frequencies.

If you are able to do so, it may be possible to modify your HF radio to allow it to transmit any mode on 136 or 472kHz at very low power. Note that modifications to a commercial transceiver, unless carried out by the manufacturer or dealer, risk seriously damaging it and will certainly invalidate any guarantee.

An approach that achieves the same end without risking damage to your rig is to use it on an HF band and transvert down to LF/MF. Apart from being able to tune the entire band, this will enable the transceiver keying circuits to be used. Transceive operation is possible if a receive converter is incorporated.

To drive a transverter your transmitter must have a milliwatt level output. This can be from a dedicated transverter port or achieved by running the lowest possible power and using an external attenuator.

G3YMC transmit/receive converter for 136kHz
The transverter described here was built by G3YMC for use on 136kHz, and converts to and from his FT101ZD tuned to the 10MHz amateur band. This band was chosen primarily because 10MHz computer crystals are cheap and readily available, and the performance of some amateur transceivers is rather better on this band than on the more commonly used 28MHz band.

No originality is claimed for the design. It was built up using published circuits in the *ARRL Handbook* and elsewhere, modified ad hoc for the application. The circuit is described in blocks, with all components shown but no detail is given on the inductors and transformers except for general guidelines. It is intended as an experimenter's unit rather than something available off the shelf. A PA is included, producing a few tens of watts, but the earlier stages can be used in conjunction with an existing power amplifier.

LF TODAY

A 10MHz JFET crystal oscillator is used. This is a conventional circuit (**Fig 5.6**) and outputs a reference to the receive and transmit mixers. A trimmer capacitor can be used to set the frequency accurately. The oscillator is powered from a stabilised 5V rail provided by a 7805 regulator to ensure stability. The two output capacitors may be adjusted in value to obtain the correct injection levels.

The receive converter (**Fig 5.7**) uses a dual gate mosfet mixer preceded by an FET preamp. Input from the antenna is applied to a fixed tuned stage, T1. This transformer uses a cut down long-wave ferrite rod from a scrap radio and resonates in the 136 band with a parallel capacitor of around 300pF. This is top capacitive coupled to a second tuned stage (using a pot core) which can be tuned from 65-180kHz with a dual 500pF variable capacitor. An input is provided here for a wideband reception antenna, but at lower intermodulation performance.

A 3N201 dual gate mosfet is used as the receive mixer, with the oscillator injection applied to G2. This is conventional with the drain resonated to 10MHz by T3, wound on a quarter inch coil former.

Fig 5.6: The 10MHz oscillator forms the heart of G3YMC's transverter

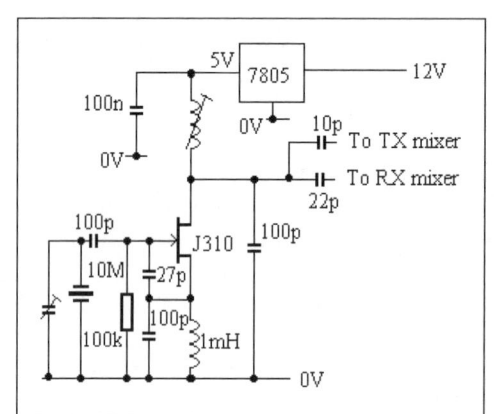

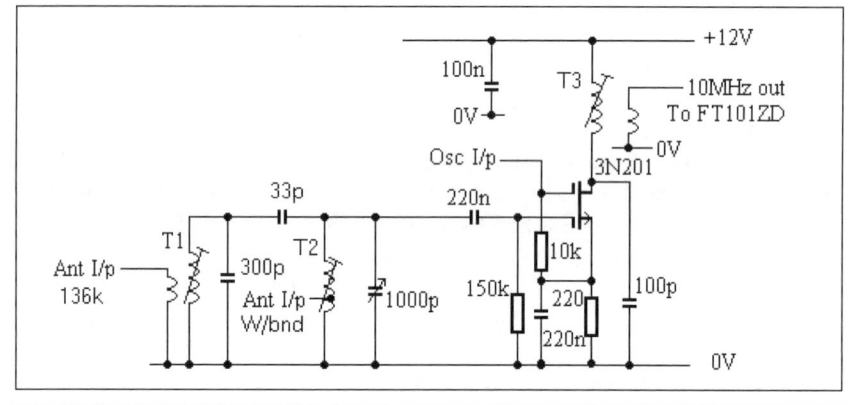

Fig 5.7: The receive converter of G3YMC's LF transverter

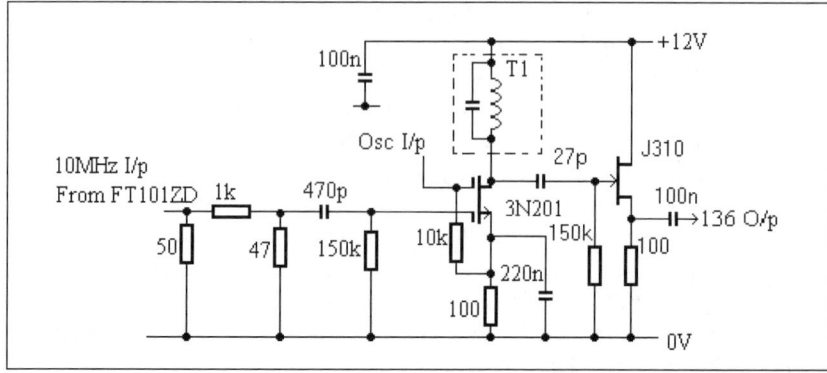

Fig 5.8: The transmit mixer

CHAPTER 5: GENERATING A SIGNAL

Fig 5.9: Transmit driver of the transverter

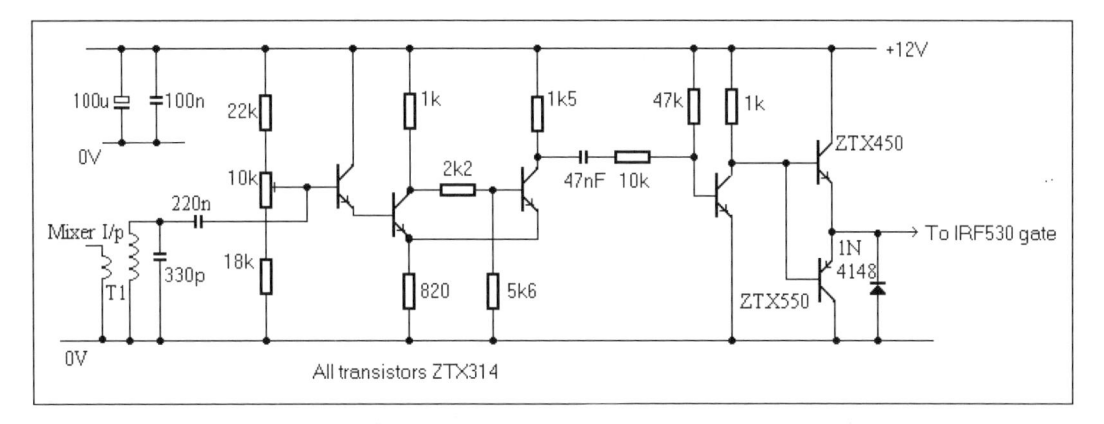

The transmit section of the transverter consists of a mixer, low level driver stage, mosfet power stage and output matching and filtering. The transmit mixer is mounted in the same box as the receive converter; the other stages are in a separate box. A 10MHz drive signal is taken from the low power output socket of the HF transceiver and fed to the transmit mixer (**Fig 5.8**) where it is terminated in 50 ohms and attenuated to a suitable level for applying to the gate of a dual gate mosfet. An injection signal is applied to the second gate from the crystal oscillator. The drain is connected to a parallel tuned circuit at 136kHz, whose output is buffered by an FET source follower and fed to the transmit driver stage (**Fig 5.9**).

The low level input from the transmit mixer is passed through a 136kHz tuned stage and then fed into a Schmitt trigger to produce an approximately square waveform. Fine adjustment of the Schmitt threshold 10k potentiometer allows the mark-space ratio to be optimised.

The drive is buffered by dual emitter followers and applied to the gate of the power mosfet. These low level stages are powered by a 12V rail derived from the 24V PA rail via a 7812 regulator. The output amplifier (**Fig 5.10**) consists of a single IRF530 mosfet run off a 24V supply. The current taken and the output power is determined by the impedance matching, which transforms the 50 ohm load impedance to around 3 ohms at the drain of the FET.

T2 and T3 are bifilar wound transformers on 3C85 ferrite rings [Note: 3C85 has now been discontinued. 3C90 is a suitable substitute - see Appendix 2 for details], each transformer being configured for an impedance transformation of

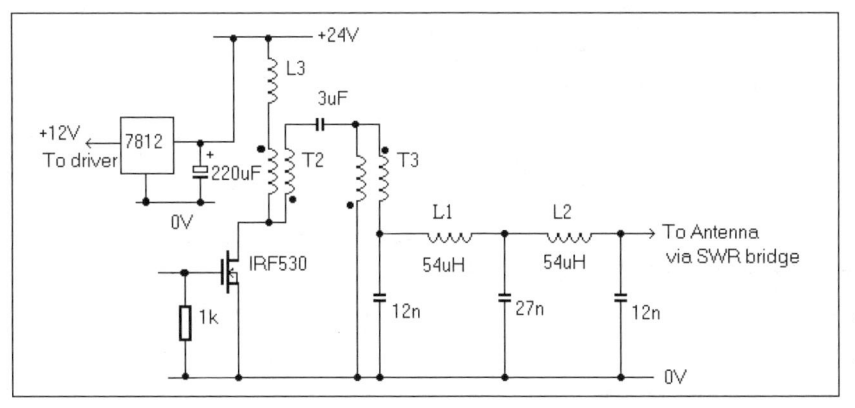

Fig 5.10: Transverter power amplifier and low pass filter

Fig 5.11: SWR monitor for the G3YMC transverter

4:1. Harmonics are reduced by a low pass filter. The inductors are 13 turns of wire on a 25mm ferrite ring.

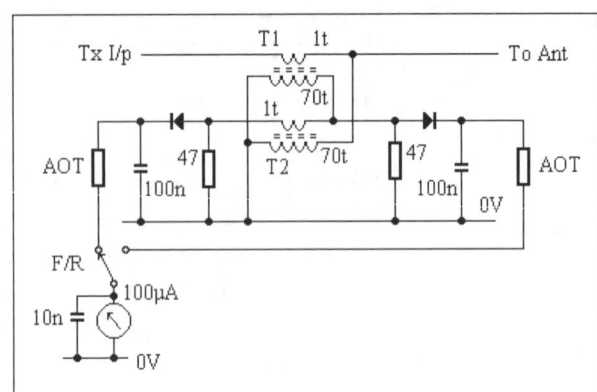

The transmitter has an output of around 35 watts into a resistive load. When using a real antenna the output may be different. A power supply of 24V at 4A maximum is recommended.

The simple SWR/Power meter shown in **Fig 5.11** can be used as a tuning aid. The value of the AOT resistors depends on the sensitivity of the meter, and is typically about 10kohms - adjust for full scale deflection in the forward direction. T1 and T2 are wound on the same 3C90 toroids as used in the transmitter. The diodes should be point contact or germanium ones with low forward voltage drop (eg OA91).

G3XBM transmit converter for 472kHz

This transverter in **Fig 5.12** was designed to work with the FT-817 working in split mode with the transmitter on 3.672 - 3.679MHz and the receiver on 472 - 479kHz. Only the transmitter needs to be down-converted. Other transceivers could be used provided the appropriate drive power is available.

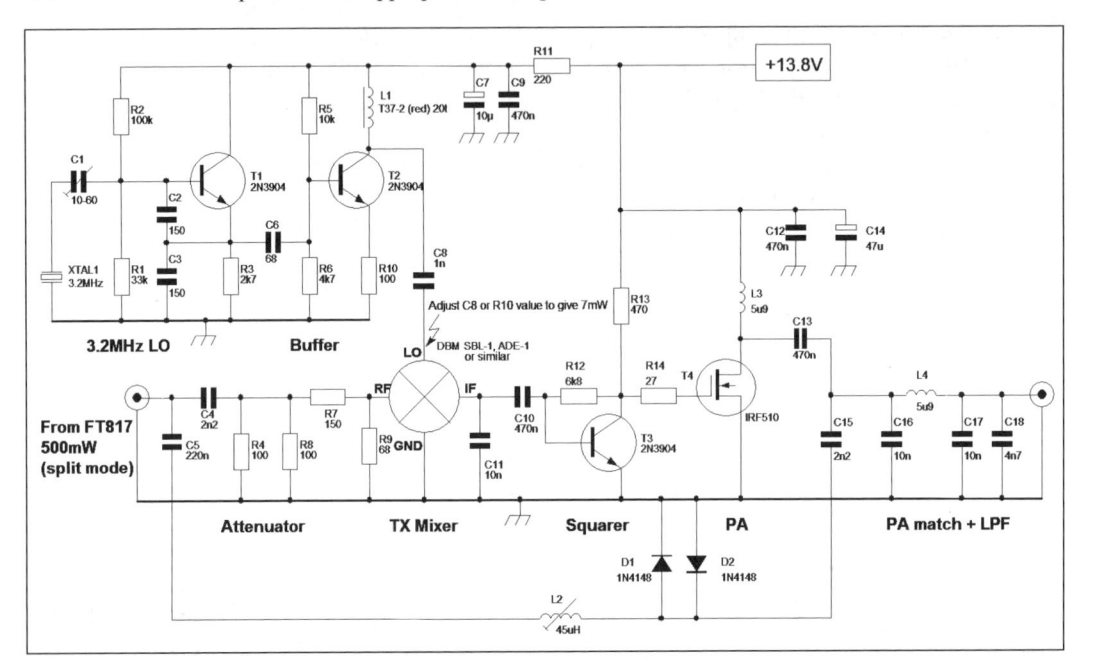

Fig 5.12: G3XBM's transmit converter

This neat version of the G3XBM 472kHz transverter was built by M1GEO

Note that L2 and C5 form a series resonant circuit on 475kHz. On transmit, D1 and D2 conduct and C15 forms part of the low-pass filter. L2 is a Toko KANK3333 or a Spectrum Communications 45u0L. On an SBL1 pin 1 is RF, 3 and 4 are IF and pin 8 is the LO; on an ADE-1 pin 3 is RF, 2 is IF, 6 is LO. All other pins on both devices are ground. A metal box is recommended for best thermal stability.

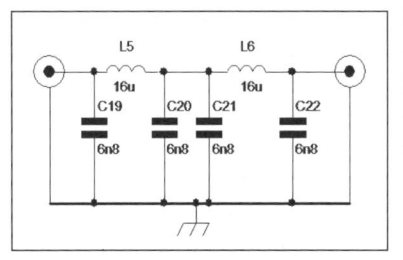

Fig 5.13: Optional external low pass filter for the G3XBM 472kHz transverter

The transverter can be used as a stand-alone MF transmitter, producing 10-15 watts RF output from a 13.8V supply, or combined with an amplifier (see the Transmitters chapter) for higher power. If required, a suitable external low-pass filter is shown in **Fig 5.13**.

G3XBM reports that results on 472kHz have been very encouraging with WSPR reports from many stations in Western Europe as far as Finland with a radiated power of less than 20mW.

Direct digital synthesis

The most sophisticated way to generate a signal at LF is Direct Digital Synthesis (DDS). A sine wave is produced from a crystal oscillator reference by entirely digital means. The DDS provides the accuracy and stability of the crystal reference, together with high tuning resolution. The main limitation of the DDS, that the output frequency is restricted to less than half the reference frequency, is not a problem for LF/MF signal generation. The DDS must be carefully designed to avoid noise sidebands and spurious emissions. Further information on this technique is contained in [2].

It is beyond the scope of this book to describe how to build your own direct digital synthesiser, but there are several circuits and kits available. Most use software stored on a chip, which relies on the constructor being equipped to do the programming or being prepared to buy the programmed chip. All require skill and experience [3, 4].

ZL1BPU's LF DDS/exciter, is not difficult to construct and has many useful facilities, including: resolution of better than 0.1Hz; high stability; support for CW, FSK, Hell and MFSK; beacon mode, sweep generator and RF output up to 1W. Full details, including how to buy the software can be seen at [5].

For the more expert contructor, DDS modules are available cheaply on eBay - just search for "DDS module".

GPS locking

Some modes used at LF/MF, especially for DX working, require very high oscillator stability and/or accurate timing. A convenient way to achieve this is to synchronise the oscillator with pulses derived from the GPS system. By using GPS modules available cheaply on the surplus market, together with some external circuitry, exceptional results can be achieved. Full details and sample projects can be found on G4JNT's web site [6].

Ultimate2 QRSS/WSPR kit

Hans Summers, G0UPL, markets a very reasonably priced kit for a DDS-controlled driver/transmitter running 100mW or so on any frequency from 500Hz to over 40MHz. It is intended as a low power beacon transmitter with pre-programmed messages being transmitted in various built-in modes including QRSS, DFCW, CW, FSK-CW and WSPR. Facilities are included for GPS locking if required. A range of inexpensive low pass filters are available from the same source [7].

The kit would make a good driver for an LF or MF transmitter, especially for beacon-only modes such as WSPR. By adjusting the settings appropriately, it can transmit an unmodulated constant carrier, which can turn it into a useful signal generator, or even as the driver for a keyed CW transmitter such as those described later in this book.

References

[1] *http://www.gw3uep.ukfsn.org/100W_QTX/472_kHz_vfo.htm* - GW3UEP 472kHz VFO

[2] *Radio Communication Handbook*, RSGB.

[3] PIC Controlled DDS VFO, 0 to 6MHz, by Johan Bodin, SM6LKM: The original ink to SM6LKM's website was not working at the time of writing, but the following links provide some information on this project: *http://www.g0mrf.com/dds.htm, http://www.njqrp.org/ham-pic/sm6lkm/*

[4] A software based DDS for 137kHz. *http://wireless.org.uk/swdds.htm*

[5] ZL1BPU's DDS/exciter. *http://www.qsl.net/zl1bpu/MICRO/EXCITER /Index.htm*

[6] *http://www.g4jnt.com*

[7] *http://www.hanssummers.com*

6

Transmitters

In this chapter:

- Ready-built and kit options
- Second-hand equipment
- Modifying audio amplifiers
- Class D transmitters
- Class E transmitters
- G3YXM 136kHz 1kW Tx
- G0MRF 136kHz 300W Tx
- Low pass filter for 136kHz
- GW3UEP's simple MF Txs
- G4JNT high power MF amp
- Low pass filter for 472kHz
- G4JNT medium power linear
- Outline of EER transverter
- Power supplies

MOST NEWCOMERS TO THE LOW frequencies will choose to start on the 472kHz band. It is much easier to make contacts on this band, because activity is greater and lower RF power is needed achieve a decent radiated signal from a practical antenna.

On the 136kHz band, antennas are much less efficient so low power transmitters are very uncommon. Experiments with QRSS have shown that it is possible to achieve ranges of several hundred kilometres whilst running only a few watts of RF (tens of milliwatts ERP), and this can be a good way to start. However, for CW and long haul contacts much more power is required. Typically, it can take several hundred watts (perhaps over 1kW) of RF to get anywhere near the 1W ERP licence limit. The exact figure depends on the size of antenna used, but a transmitter delivering more than 100 watts is very desirable.

The 136kHz and 472kHz bands in the UK (and many other countries) have a limit on radiated power and not on transmitter power. Therefore it is possible to compensate for an inefficient antenna simply by increasing the RF fed to it. In practice, however, on smaller antennas the higher voltages (or currents in the case of loops) eventually lead to a practical limit on the amount of power that can be used. So the first step that should be taken is to make the antenna as efficient as possible (see antennas chapter).

Ready-built and kit options

At the time of writing, commercial multiband HF transceivers do not include the 472kHz band. Even if this eventually becomes commonplace, it is unlikely that any will add 136kHz. Most operators build their own equipment, and they often report this to be part of the challenge, enjoyment and satisfaction. Nevertheless there is still some hope for those who are unable or unwilling to build their transmitters from scratch.

LF TODAY

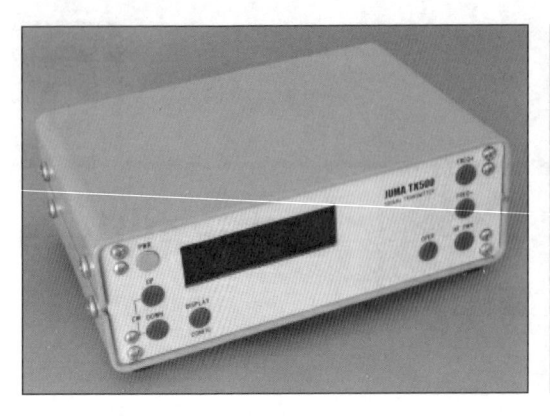

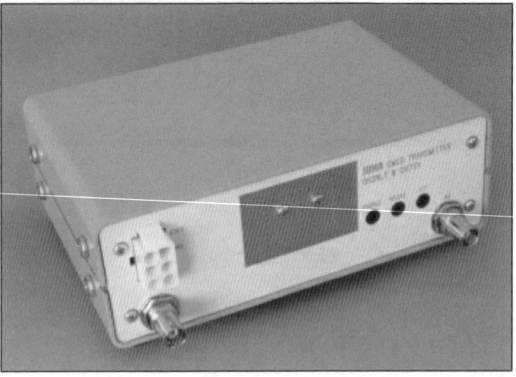

The Finnish-made Juma500 is a 60 watt transmitter for 472kHz, with a receive converter and many other built-in features. A 136kHz version is also available.

Although a few data modes require linear amplification (see the chapter on Modes), most low frequency transmitters use high efficiency non-linear class D output stages.

Juma kits for 472kHz and 136kHz

The Juma500 and Juma136 are a pair of professional looking single-band kits from Finland with an almost identical specification. The class D transmitters will run up to 60W continuous RF from 14 volts DC with SWR protection. The kits cover 450-550kHz and 130-150kHz respectively in 10Hz steps (1Hz steps for very slow QRSS is available with the supplied Windows software).

Transmit-receive switching is included, as is a receive preamplifier. Also built-in is a receive converter with an output on the 3.5MHz band. Another included feature is a keyer with sidetone. The keyer will produce Morse at 1 - 50WPM, and will send QRSS with the Windows software. There's also a beacon facility.

All of that comes at a price - the Juma transmitters are not cheap, especially when shipping and import duty is taken into account - but they will provide everything you need to operate on each of the the low frequency bands. All you need to add is an antenna and a power supply. Note that although 60W is adequate for 472kHz, an external amplifier or very large antenna may be needed to achieve success on the 136kHz band. More details can be found at [1].

TX-2200 for 136kHz

This is a 100 watt ready-built transmitter made by Japanese company Thamway. It covers 135.7 to 136.8kHz in 10Hz steps, and includes transmit-receive switching. Stability is quoted as 50ppm. An optional high stability oscillator achieves 2.5ppm, can be tuned in 1Hz steps. A further optional extra includes a frequency shift for DFCW and better heat sinking for QRSS use. The TX-2200 is expensive but is available from UK distributors [2]. More information on the TX-2200 transmitter can be found at [3].

Second hand / surplus equipment for 136/472kHz

The First
At the end of the 1990s, a Dutch company, Ropex, produced the first commercial 136kHz amateur band transmitter, which they called 'The First'. It was sold ready built, ran up to 130W of crystal controlled CW from a 12V supply and included keying and antenna changeover circuits. Very few were made, but one may occasionally be available secondhand. A review can be found at [4].

Aircraft beacons
Occasionally, transmitters suitable for modification for use on the 472kHz band become available on the second-hand market. These were originally used as non-directional beacons (NDBs) at airports. They are usually heavy duty rack-mounted crystal controlled transmitters, designed to run 100 watts or more of continuous MCW.

Some surplus equipment for 136kHz may be found on the second-hand market. The most popular was built by Racal Decca for the Decca Navigation System that was scrapped at the end of the 20th century [5]. The transmitter units, which operated in the 70 - 130kHz region, are on very bulky steel frames designed for rack-mounting, but are metered and well protected. They can run up to 1.2kW and require two power sources, 24V at 2 amps, and between 50 and 80 volts at up to 25 amps. Some of them work with almost no modification, whilst others - the lower frequency versions - require a small amount of tweaking [6]. The Decca units contain only the driver and output stages and must be fed from a 136kHz signal source, such as a DDS - see previous chapter.

Ex-Decca Navigation LF transmitter, showing the three transmitter modules at the top. Most of the chassis space is occupied by the tank circuits

Modifying audio amplifiers for 136kHz
A few audio amplifiers can be made to work at 136kHz. Hi-fi amplifier units with onboard mains power units are available from BK Electronics [7] giving a few hundred watts of audio.

These amplifiers can provide up to 100W at 136kHz and at around £100 they make an inexpensive and readily available starting point for those who are

The BK Electronics 300-watt audio amplifier is capable of 100 watts or so at 136kHz

unwilling to build their first LF transmitter from scratch, but who are still prepared to do a few modifications [8] to make the equipment work way beyond its design frequency.

The Hafler P3000 audio amplifier has been used in the USA. It runs up to 300 watts and its bandwidth is specified as 300kHz. This makes it suitable for linear operation on 136kHz, and therefore all data modes, though it is rather expensive. A data sheet can be found at [9] and the experiences of W1TAG, whose 136kHz beacon WD2XES is frequently received in Europe, is at [10].

Although there is some old marine equipment available for 500kHz, the low power requirements of operating at this frequency make it simple to build your own transmitter.

Home made low frequency transmitters

Most signals on the 136 and 472kHz bands come from home-made transmitters. It is easier to construct transmitters for 472kHz because these usually involve lower power than is typical on 136kHz.

In the previous chapter we looked at various ways of generating a small signal on the required band. The following pages describe several transmitters and amplifiers that produce output power levels needed for practical operation.

Although there are some modes that require linear amplification, most can use relatively simple Class D (or E) amplifiers. These employ power mosfets in push-pull, sometimes with several transistors in parallel. A number of these feature in this chapter. Some constructional experience is required, but there is always support for the constructor, both from the original designers and from those amateurs who have already built the circuits (see the section on further information).

If higher power is required, two power amplifiers can be 'paralleled' using a Wilkinson Combiner [11].

Class D transmitters - design notes

Low frequency transmitters are usually required to produce between 100W and a few kilowatts of output. Most LF operators are currently using class D switching-mode output stages; these can achieve very good efficiency, which considerably simplifies cooling problems associated with high-power linear amplifiers.

These circuits are also well suited to inexpensive power mosfets and other components intended for switch-mode power supplies operating in a similar frequency range.

Switching mode circuits are, however, more difficult to key or modulate satisfactorily. Fortunately, most LF operation uses simple on-off keying or frequency-shifting.

Class-D amplifiers fall into two distinct types: voltage-switching, **Fig 6.1(a)**; or current-switching, **Fig 6.1(b)**. In each case, the load is connected to the output stage via a resonant tank circuit.

The voltage-switching type has a series-tuned tank circuit; the switching mosfets develop a square-wave voltage at the input side of the series tank circuit. However the tank circuit ensures the current flowing in the load is almost a pure sine wave.

CHAPTER 6: TRANSMITTERS

Fig 6.1: (a) Voltage-switching class-D amplifier, and (b) Current-switching class-D amplifier with mosfet drain waveforms

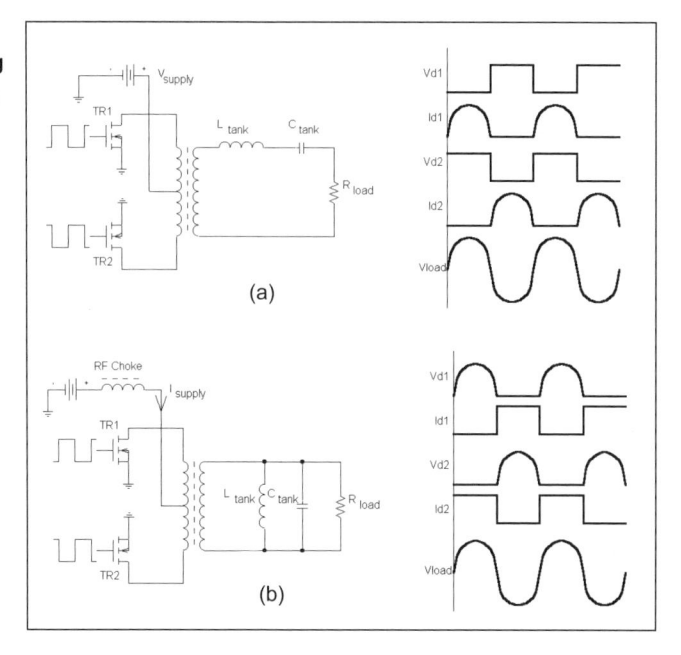

The current-switching type has a parallel-tuned tank circuit; the supply to the output devices is a constant current which is applied to the tank circuit in alternate directions depending on which mosfet is switched on. The resulting square wave current applied to the tank circuit again results in an almost sinusoidal voltage across the load.

Since a constant-current DC supply is not very practical, a constant voltage supply is used with a series RF choke. Provided the impedance of the choke is much greater than the load resistance, the supply current is almost constant. The major advantage of class D is that the mosfets are either fully 'on', in which case the only power loss is due to the mosfet 'on' resistance, $r_{DS(on)}$, or fully off, with essentially zero power dissipation. In practice, there are additional losses, but these are small compared to linear amplifiers, and efficiency can exceed 90%.

In many amateur circuits, the tank circuit is replaced by a low-Q low-pass filter, **Fig 6.2**. This circuit is 'quasi-parallel resonant'; it provides a resistive load at the output frequency, but a low shunt impedance at the harmonics. The low Q leads to non-ideal class D operation in that the voltage waveform is not a perfect

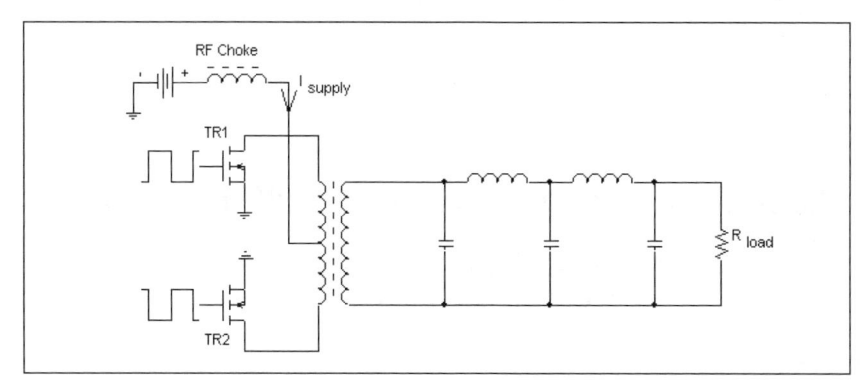

Fig 6.2: Practical form of Class D output stage

Fig 6.3: Class D waveforms. Upper trace, drain voltage; middle trace, drain current; lower trace, gate drive voltage

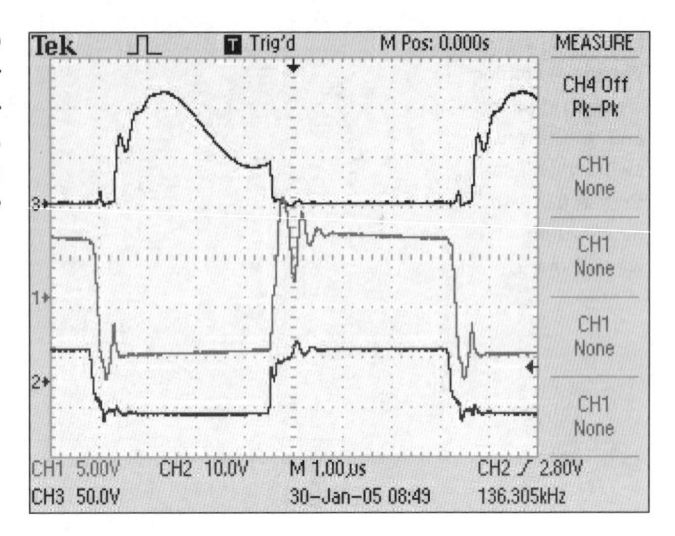

sine wave, but has the advantages that smaller inductors and capacitors are required, tolerances are less critical, and better rejection of higher harmonics is provided by the multiple filter sections. The voltage and current waveforms of a real-world class D output stage using this circuit are shown in **Fig 6.7** Compared to the idealised waveforms, some high frequency 'ringing' is visible. This is due to stray capacitance and inductance which inevitably exists in the circuit, and is undesirable since it causes increased losses, as well as the potential for generating high-order harmonics. It is therefore important to minimise stray reactance, two important causes of which are the parasitic capacitance of the mosfets themselves, and the leakage inductance of the output transformer. Adding damping RC 'snubber' networks can also usefully reduce the level of ringing.

As with other types of amplifier, the output power of class D amplifiers is defined by the supply voltage V_{cc}, output transformer turns ration n and load impedance R_L. For the voltage switching amplifier:

$$P_L = \frac{8n^2 V_{cc}^2}{\pi^2 R_L}$$

While for the current-switching class D:

$$P_L = \frac{\pi^2 n^2 V_{cc}^2}{8 R_L}$$

These formulas assume losses in the circuit are negligible; in practice, some losses do occur but since they are small, the results given by the formulas are reasonably accurate.

Class D PA design example - 200W on 136kHz

The design process for a class D transmitter output stage is best illustrated by an example. The following design is for a 136kHz transmitter with about 200W output, using a current-switching class D circuit.

This is a modest power level for 136kHz, but the principles discussed have been applied equally well to designs with 1kW or more output using this circuit

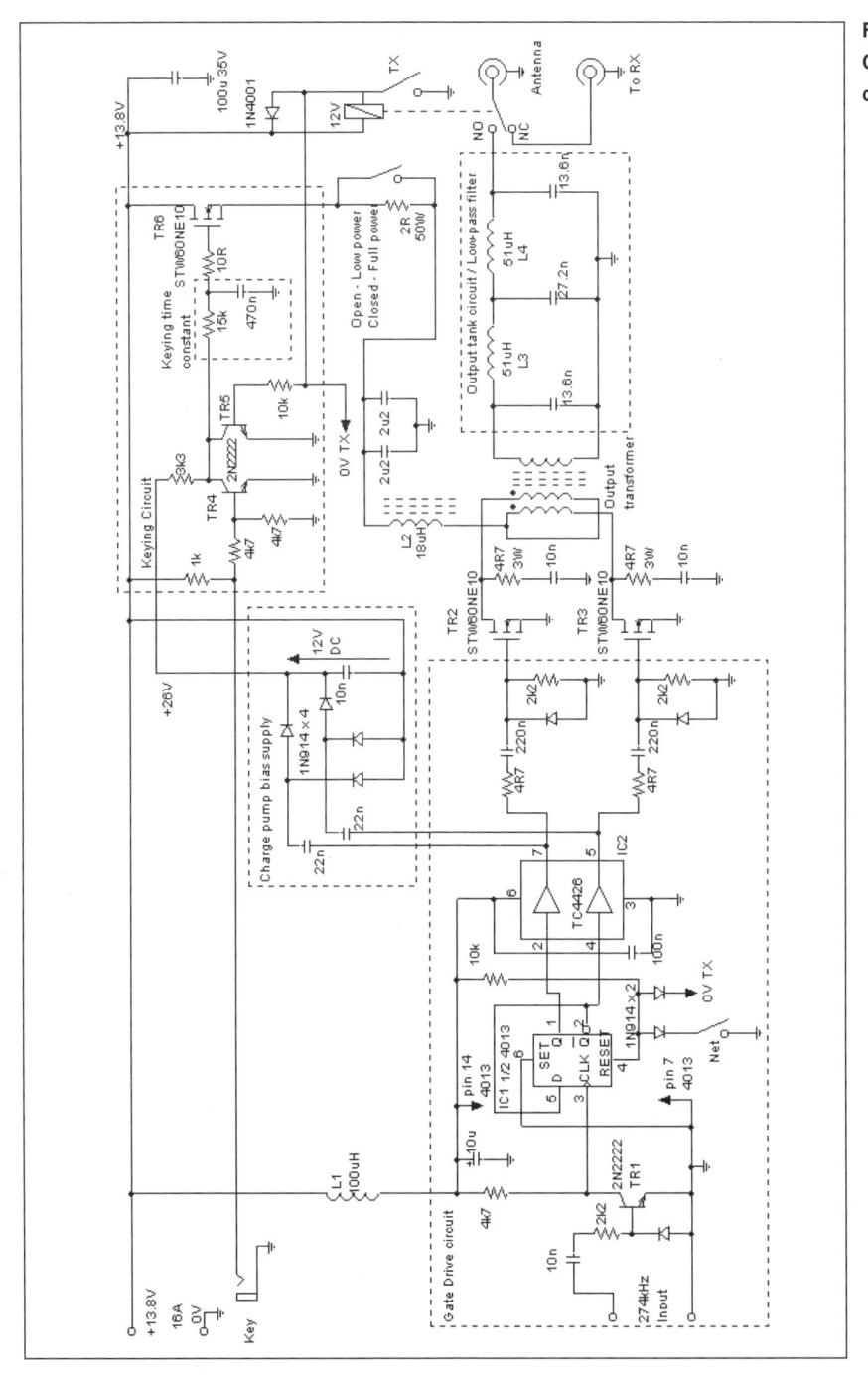

Fig 6.4: 200W Class D transmitter circuit

configuration, which is probably the most popular in use at present. The complete circuit is shown in **Fig 6.4**.

The first design decision is what DC supply voltage to use, since the power supply is normally the most expensive and bulky part. In this case 13.8 volts was selected; it can use the standard DC supply found in many amateur shacks.

The DC input power required will be about 10% greater than the RF output due to losses, so the expected supply current will be 220W / 13.8V = 16A, a level that most 13.8V supplies can readily deliver. For higher power designs, 40 to 60V is often a good compromise, since the problem of large DC and RF currents is then reduced. It is perfectly possible to use an 'off line' directly rectified AC mains supply with no bulky mains transformer, as has been done by G4JNT [12]; note that design for *electrical safety is absolutely critical in this case*. Inexpensive switching mosfets are available suitable for any of these supply voltages.

In the ideal push-pull current switching circuit, the peak mosfet drain-source voltage will theoretically be π times the DC supply voltage, in practice about four times the 13.8VDC supply is likely. Mosfets should be selected so that only a few percent of the DC input power will be dissipated in their 'on' resistance, $r_{DS(on)}$. This condition also ensures the mosfet will have adequate drain current rating. STW60NE10 devices were used for TR2, TR3, with BVDS of 100V, and a typical $r_{DS(on)}$ of 0.016 ohms, leading to about 4W dissipation due to the 'on' resistance and 16A supply current ($I^2 \times r_{DS(on)}$).

Additional dissipation occurs during the transient period where the device is switching 'on' or 'off'. This can be determined from measurement of circuit waveforms, but can be assumed to be similar to that due to $r_{DS(on)}$. In normal operation therefore, each mosfet will only dissipate a few watts; however with a severe mismatch, power dissipation can be much higher, especially without DC supply current limiting. For a robust design, the mosfets and their heatsink should be able to dissipate of the order of 50% of the total DC input power, at least during a short overload period. The STW60NE10 devices have a TO-247 package and can dissipate 90W each at a case temperature of 100 degrees C, which is adequate.

The output transformer is the most important part of the design. It is normally wound on a core using the same ferrite grades that are used for switch-mode power supplies. These may be large toroidal cores, pot cores, or 'E' cores with plastic bobbins.

Several manufacturers produce suitable materials; these include Ferroxcube (Philips) 3C8, 3C85, 3C90, Siemens N27, N87, Neosid F44, and Fair-Rite #77 grades. All these ferrites have permeability around 2000, and have reasonably low loss at 137kHz. They are available in a variety of forms, such as EE, EC, and ETD styles, and sizes; a designation such as ETD49 means an ETD style core that is 49mm wide.

A good selection of different core types is available from component distributors [13, 14] at reasonably low cost.

Transformer design is a complex topic in its own right, but a simplified procedure usually gives satisfactory results for amateur purposes, as follows.

Given the supply voltage and load impedance (usually 50 ohms), the turns ratio of the output transformer determines the output power. Rearranging the formula for current-switching class D given in the previous section gives:

$$n = \sqrt{\frac{8 P_L R_L}{\pi^2 V_{cc}^2}}$$

For V_{cc} = 13.8V, R_L = 50 ohms, and P_L = 220 ohms, this gives n = 1:6.8. Next, a suitable sized core is chosen. As a guide, using the types of ferrite listed above, an ETD34 core is suitable for powers up to 250 watts, an ETD44 core for 500 watts, and an ETD49 for up to 1kW. Similar sizes in different styles have similar power handling. If in doubt, use a bigger core!

The number of turns N in the secondary winding can then be determined. N must be large enough to keep the peak magnetic flux B_{peak} to a value well below the saturation level at the expected output voltage level, V_{RMS}:

$$B_{peak} = \frac{V_{RMS}}{4.44fNA_e}, \quad V_{RMS} = \sqrt{P_L R_L}$$

Where A_e is the effective area of the core in m^2. The number of turns must also be large enough so that the inductance of the winding has a large reactance compared to the load impedance. A value of X_L about 5 - 10 times the load impedance is desirable. An ETD34 core and bobbin of 3C85 ferrite material was available, which according to the manufacturer's data has A_e of 97.1mm2 (97.1 x 10^{-6} m^2), and A_L of 2500nH/T^2.

A suitable maximum value of B_{peak} for power-grade ferrite materials is around 0.15 tesla. For 220W output, V_{RMS} is 105V A few trials using the formulas resulted in n = 14 turns, B_{peak} = 0.127T and L = 490uH, X_L = 422 ohms, which meets the criteria given.

Two turn primary windings result in a turns ratio of 1:7, close enough in practice to the 1:6.8 design value. The primary windings were 4 x 2 turns, quadrifilar wound of 1mm enamelled copper wire, using two windings in parallel for each half of the primary winding. The secondary of 14 turns, 0.8mm enamelled copper was wound on top of the primaries, and insulated from them with polyester tape.

The DC feed choke L2 must be capable of handling the full DC supply current without saturation and also have a high reactance at 137kHz compared to the load impedance at the transformer primary, which is (50Ω / 7^2), about 1Ω. A reactance of 10Ω or greater is adequate, requiring at least 12uH. A high Q is not required, since only a small RF current flows in the choke. The 18uH choke used a Micrometals T-106-26 iron dust core. Iron dust cores of similar types to this can often be salvaged from defunct PC switch-mode PSUs. The winding used 2 x 17 turns in parallel of 1mm^2 enamelled copper wire. An air-cored inductor would also be feasible, if more bulky.

The output filter consists of two identical cascaded pi-sections. The filter should provide a resistive load at the 137kHz output frequency, but a low capacitive reactance at harmonics. This can be achieved by designing the pi-sections as low-Q matching networks, with equal source and load resistances. This yields a circuit with two equal capacitors. The standard pi-section design formulae can be used, modified for $R_{in} = R_{out} = R$:

$$X_C = \frac{R}{Q}, \quad X_L = \frac{2QR}{Q^2+1}$$

$$C = \frac{1}{2\pi f X_C}, \quad L = \frac{X_L}{2\pi f}$$

Most designers select Q between 0.5 and 1. Metallised polypropylene capacitors are a good choice, since they have low losses at 137kHz, and are available with large values and high voltage ratings. The DC voltage rating should be several times larger than the RMS RF voltage present; the main limitation is the heating effect of the RF current causing internal heating of the capacitor. Several 6.8nF, 1kV polypropylene capacitors were available, so C = 2 x 6.8nF = 13.6nF was used. This has reactance of 85.4Ω at 137kHz, forcing Q = 0.585, and giving XL = 43.6Ω, L = 50.6uH.

The inductors must have low loss at 137kHz to avoid excessive heating. Micrometals T130-2 iron dust cores were used, wound with 68 turns of 0.7mm enamelled wire. It is a good idea to check the capacitance and inductance of the filter components using an LCR meter or bridge. However, the main effect of small errors is only to slightly alter the output power from the circuit, without greatly affecting the efficiency.

The drive signal applied to the class D output stage is a 50% duty cycle square wave. The 137kHz gate drive signal is obtained from a 274kHz input using a D-type flip-flop in a divide-by-two configuration, guaranteeing an accurate 50% duty cycle. When the circuit is switched to receive, the flip-flop is disabled by pulling the reset input high, preventing a 137kHz signal leaking to the receiver input and causing interference. For netting the transmitter, the 'net' switch enables the flip-flop.

The mosfets require zero or negative gate voltage to switch the transistor fully off, and +10 to +15 volts to bias them fully on. The mosfet gates behave essentially as capacitors, requiring transient charging and discharging currents as the drive voltage switches on and off, but drawing no current while the gate voltage remains stable. In order to achieve fast mosfet switching, a TC4426 gate driver IC is used. These driver ICs accept a TTL-compatible logic level input signal, and are designed to produce peak output currents of 1A or more which charge and discharge the mosfet gate in a fraction of a microsecond. A disadvantage of using a flip-flop to generate the drive signal is that if the input signal is lost one mosfet will remain switched on and act as a virtual short across the supply. To avoid this, the gate driver is capacitively coupled to the mosfets; the shunt diodes perform a DC restoration function, making the full positive peak voltage available to drive the gate. If drive is lost, the gates discharge through the 2.2kΩ resistors, switching both mosfets off. The 4.7Ω resistors in series with the gate drivers help to reduce ringing.

Each mosfet has a series RC damping network from drain to source, reducing high frequency 'ringing' superimposed on the drain waveform. The component values are best determined experimentally, since they depend on the individual circuit. A good starting point is to make the capacitor about five times larger than the mosfet output capacitance. A resistance between 2 and 20 ohms is usually effective. Effectiveness is best checked by examining the mosfet drain waveforms with an oscilloscope, and compromising between minimising high-frequency ringing and excessive power dissipation in the resistors. Larger capacitors and smaller resistors normally result in reduced ringing, but increased dissipation. These components should be appropriate for high frequency use; in this circuit, 4.7Ω, 3W metal film resistors, and 10nF, 250V polypropylene capacitors were satisfactory.

The transmitter is keyed using series mosfet TR6. The mosfet should have low $r_{DS(on)}$ to minimise loss when switched on. A third STW60NE10 was used, although since the maximum voltage applied to this device is the 13.8V DC supply, a lower voltage device could be used instead. During the rise and fall of the keying waveform, dissipation in the keying mosfet peaks at about 25% of the maximum DC input, 55W in this case. However, when the mosfet is fully on, it dissipates only a few watts due to $r_{DS(on)}$, and when fully off dissipation is practically zero. Therefore the average power dissipated is small, under 10W in this circuit, provided it is not keyed very rapidly.

In order to bias this mosfet fully on, a voltage around 10V higher than the 13.8V DC rail must be applied to the gate. Only a few milliamps of bias is required; a small auxiliary DC supply could be used in a mains-powered transmitter, but in this case the bias voltage was obtained by rectifying the capacitively-coupled gate drive waveform using a charge-pump circuit. The bias is controlled by the key input via transistor TR4, and the keying waveform is shaped by the RC time constant to give around 10ms rise and fall times. TR5 maintains the keying 'off' when the circuit is switched to receive. The 15V zener across the mosfet gate and source limits the gate voltage to prevent damage. The output power of a class D transmitter can be varied either by changing the supply voltage, or by having multiple taps on the output winding; both these techniques are used in the G0MRF and G3YXM designs described later. In this design, a resistor can be switched in series with the DC supply, reducing the supply voltage to the output stage to around 4V, and RF output to 18W, for tuning-up purposes. The 2Ω wirewound resistor dissipates nearly 40W in low power mode, so greatly reduces efficiency, but does make it very difficult to damage the PA due to its inherent current limiting, useful when using a battery supply or when initially testing the circuit.

Construction of this type of transmitter is reasonably non-critical. The low-power parts of the transmitter can be assembled using 'veroboard' or similar, but the gate driver IC must have a 0.1uF ceramic decoupling capacitor directly across the supply pins due to the large transient currents present. Also for this reason, the gate leads, and the ground return from the mosfet sources should be kept very short (<30mm).

The mosfets, output transformer, keying circuit and DC feed carry heavy currents, so connections should be as short as possible and use thick wire (at least 2.5mm²). The RC damping components should be mounted directly across the mosfet drain and source pins, and the connections to the output transformer kept short. The circuit described above was assembled on an aluminium plate about 160 x 200 x 3mm, which provided ample heatsinking for the three mosfets when air was allowed to circulate freely. In an enclosed box, a fan would probably be desirable.

Testing a class D circuit should start by checking operation of the gate-drive circuit, ensuring that complementary 12Vp-p square waves are present at the mosfet gates at the correct frequency. A dummy load is almost obligatory for testing LF transmitters (see the Measurement and Calculation chapter). If possible, apply a reduced DC supply voltage to the output stage (but not to the gate drive circuit!), or use a series resistor to reduce the supply voltage, as included in this circuit. An oscilloscope is the ideal tool to check the correct waveforms are present. A useful check is the efficiency; the ratio of RF output power to DC input power should be well over 80% if the circuit is working correctly.

Class E transmitters

Class E power amplifiers are another form of switch-mode output stage that have been successfully applied to 136kHz and 472kHz amateur transmitter construction. The circuit was invented by Nathan Sokal, WA1HQC [15]. The basic class E circuit is single-ended, as shown in **Fig 6.5(a)**. The single switching mosfet drives a tank circuit C1, L2, C2 with series and shunt capacitors, and is fed with DC supply current via a high impedance choke L1.

As in class D PAs, the active device (typically a power mosfet) is used as a switch that is either fully 'off' or 'on' and thus, in an idealised circuit, no power is dissipated. In switch mode circuits in general, some small losses will occur, partly due to the unavoidable 'on' resistance of the mosfet, but also partly due to the finite time taken for switching to occur. During switch on, the current is rising while the voltage falls and the reverse occurs during switch off. The instantaneous power dissipation can be high during these short transition periods, since both voltage and current in the mosfet is simultaneously quite large. The class E tank circuit design aims to minimise these losses by shaping the voltage and current waveforms in the mosfet so that:

- The mosfet current decreases to zero before switch-off occurs, so the MOSFET current is zero while the voltage rises
- The voltage across the mosfet reaches zero before switch-on occurs, so current in the mosfet does not rise until the voltage is zero

The approximate waveforms of the voltage and current are shown in **Fig 6.5(b)**. The output waveform delivered to R_L is effectively filtered by the series-tuned circuit formed by L2, C2 and is an almost pure sine wave. Analysis of this circuit is complex due to its non-linear nature and the non-sinusoidal waveforms. A good description, together with detailed design formulas has been provided in [15]. An approximate design procedure is as follows. Initially, the operating frequency f, supply voltage V_{CC}, output power P_L and tank circuit loaded Q, Q_L are selected. Typically Q_L is between 5 and 10. Referring to Fig 6.5(a), the component values are approximately:

(left) Fig 6.5: (a) Basic class E PA circuit
(right) (b) Class E waveforms

$$R_L \approx 0.577 \frac{V_{CC}^2}{P_L} \;;\; C_1 \approx 0.184 \frac{1}{2\pi f R_L}$$

$$C_2 \approx \frac{1}{2\pi f Q_L R_L} \;;\; L_2 \approx \frac{Q_L R_L}{2\pi f}$$

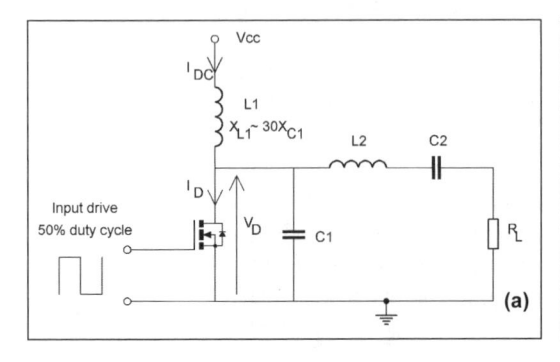

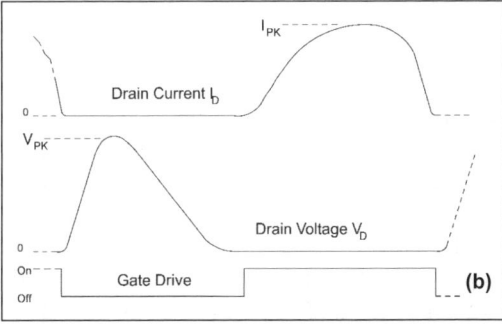

The DC current supplied to the PA, and the peak values of the voltage and current waveforms of Fig 6.5(b) are approximately given by:

$$I_{DC} \sim \frac{P_L}{V_{CC}}; \quad V_{PK} \sim 3.56 V_{CC}; \quad I_{PK} \sim 2.86 I_{DC}$$

These approximations neglect the effect of losses, finite tank circuit QL, the impedance of the feed choke L1, but yield values within typically 20%. To obtain more accurate values, the more complex formulas in [15] should be used; design spreadsheets are available from [16] to simplify this process. In any case, due to component tolerances and parasitic capacitance and inductance in the circuit, it is generally necessary to adjust the tank circuit components experimentally to their final values while observing the voltage waveform with an oscilloscope. A detailed description of this process is also given in [15]. The value of R_L will generally not be 50 ohms; a ferrite-cored matching transformer may be used to obtain the desired load impedance as for the class D circuit, or the tank circuit may be extended to include an additional L-network as described in the G4JNT design in [16, 17] and below. Additional low-pass filtering for harmonic suppression may also be required.

Compared to class D PAs, perhaps the main attractions of class E designs for amateur LF/MF use include a simpler circuit with single-ended input drive and output, and 'cleaner' waveforms with less possibility of high frequency ringing and switching transients, especially compared to the push-pull current-switching class D circuit. This is probably mostly due to the elimination of the push-pull transformer with its inevitable parasitic inductance and capacitance. At higher frequencies, switching losses in class E PAs are significantly reduced compared to class D, but at 136kHz and 472kHz, efficiencies of well over 80% are possible with both types. The main disadvantages of class E are the need for relatively bulky high Q tank circuit inductors and capacitors, and the attendant requirement for a more precise tuning up procedure.

G3YXM 136kHz 1kW transmitter

G3YXM set out to design a transmitter that is reasonably small, produces around 1kW RF output, and will withstand antenna mis-match and other mishaps. The description here is an abridged version of the original article available via G3YXM's web pages [18]. The circuit is shown in **Fig 6.6** and the major components are listed in **Table 6.1**.

An input signal at 1.36MHz from the VFO (**Fig 6.7**), or from a crystal oscillator and further divider, is divided by ten, the output at IC4 being a symmetrical square wave, driving the output mosfets via gate driver IC6. Each mosfet shown is actually two devices in parallel. The output transformer ratio is set by switch S2. Higher output is obtained with more turns selected. Across the primary of the transformer the Zobel network marked 'Z' (22Ω and 4n7 in series) reduces ringing. The output is fed to the antenna via a low-pass filter. The cut-off frequency is quite high at about 220kHz as virtually no second harmonic is produced.

The SWR bridge consists of T4 and associated components. It is a bifilar winding of 2x18 turns which forms the centre-tapped secondary and the coax inner passing through the toroid core forms the single-turn primary. The protection circuit which cuts the drive for about a second is triggered by high SWR via IC1B,

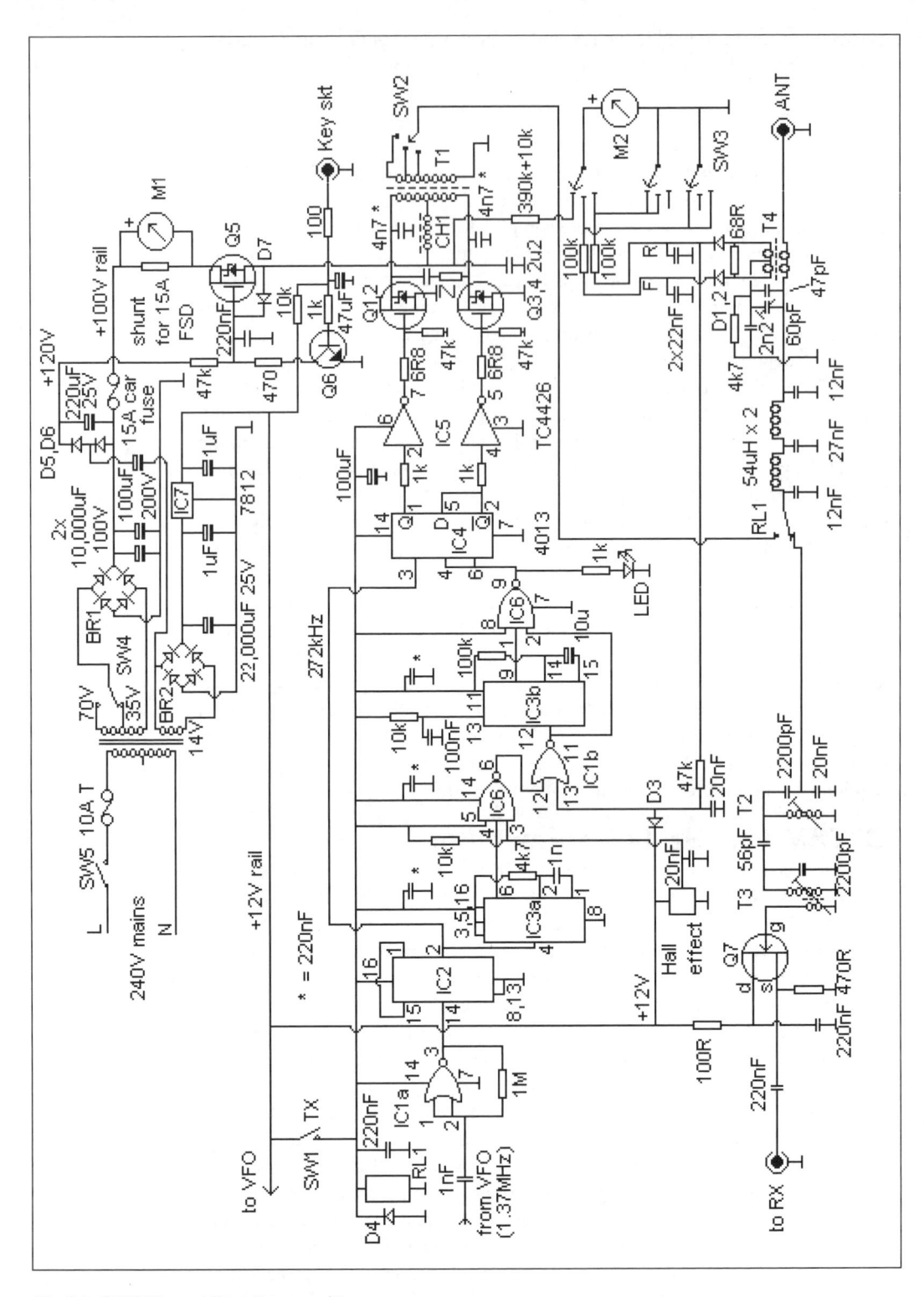

Fig 6.6: G3YXM one kilowatt transmitter

CHAPTER 6: TRANSMITTERS

IC1	HEF4001
IC2	HEF4017
IC3	HEF4538
IC4	HEF4013
IC5	TC4426
IC6	HEF4023
IC7	7812
Q1, 2, 3, 4	IRFP450
Q5	IRFP260
D1, D2	1N4936
D3, D4, D4, D6	1N4006
Hall effect device	OHN3040U (Farnell 405-656)
BR1, BR2	35A, 600V (Farnell 234-151)
R (for Zobel network "Z")	22Ω, 25W (Farnell 345-090)
Mains transformer	2 x 35V, 530W
T1	Primary 2 x 8 turns, secondary 20 turns, tapped at 12 & 16 turns
CH1	20 turns on 50mm length of antenna-type ferrite rod
T2, T3	Toko 719VXA-A017AO (Bonex)
T4	Primary 1 turn, secondary 2 x 18 turns bifilar
Output filter inductors	54µH. 65 turns 1mm enamelled wire on Micrometals T200-2 toroid core

Table 6.1: Component details for the G3YXM 1kW transmitter

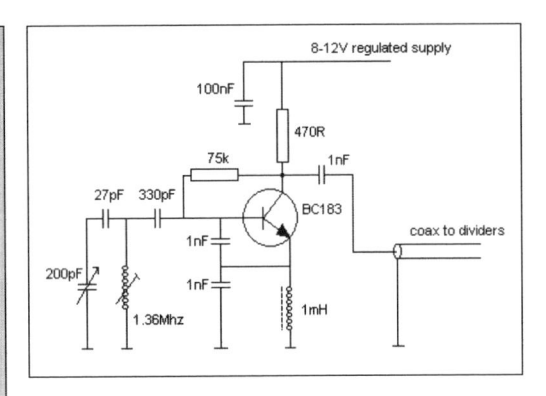

Fig 6.7: 1.36MHz VFO for the G3YXM 136kHz transmitter

or over-current signal from the Hall-effect device, which is triggered by the magnetic field of CH1, which is made from a 50mm piece of ferrite antenna rod wound with 20 turns of 1.5mm enamelled copper wire. The Hall-effect detector is placed near one end of CH1 and the spacing adjusted to trip at about 20A. The receive pre-amp uses coupled tuned circuits giving a band-pass response over the 135 to 138kHz range A single Jfet (Q7) makes up for the filter loss.

The mains transformer used in the power supply has two series 35V windings. The DC voltage is either 50V from the centre tap or 100V from both windings. An auxiliary 12V winding was added by winding 30 turns of 16SWG wire through the toroid. At full output the HT will drop to about 80V. The keying circuit uses a series mosfet with shaping of rise and fall times to prevent key-clicks. To turn this mosfet fully on, its gate must be at least 5V positive of its source which is close to the main supply voltage. Diodes D5 and 6 are supplied via a high voltage capacitor from the 12V winding to produce an extra 20V bias for this purpose.

The low-level circuitry was built on strip-board, taking care to keep the tracks short and earth unused inputs. The TC4426 chip IC5 is capable of driving 1.5A into the gate capacitances of the mosfets and the decoupling capacitors must be fitted close to the chip with short leads. The 6R8 series resistors are mounted on the gate pins of the mosfets, the resistor leads forming the connections to the strip board. It is probably best to use one resistor for each gate. The strip board should be grounded to the earth plane as near as possible to the mosfets, which should have the source leads soldered to the ground plane. The two 4n7 capacitors should be connected directly across the mosfets.

Output transformer T1 is constructed from two-core 'figure of eight' speaker cable wound eight times through the ferrite toroid, connected as a centre-tapped primary by connecting one end of one winding to the opposite end of the other. The secondary is wound over it with 20 turns of thin wire tapped at 12 and 16 turns. The Zobel network should be wired from drain to drain with short wires.

Get the PSU, VFO and CMOS stages working first. Check with a scope that you have complementary 12V square waves on the gates, the waveform will be slightly rounded off due to the gate capacitance. Connect the transmitter to a 50Ω load and, having selected the first tap on SW2, apply 50V (SW4 in low position) with a resistor in place of the fuse to limit the current. The mosfets should draw no current without drive.

Press the key and the output stage should draw a few amps and produce a few watts into the dummy load. If the shut-down LED comes on, either the load is mis-matched, the SWR bridge is connected backwards or the 60pF capacitor needs adjustment. If all seems well, remove the current limiting resistor and increase the power by selecting taps, key the rig in short bursts and check for overheating of mosfets and cores. When you are happy that the transmitter is working OK, load it up to 15A PA current and slowly move the Hall device nearer to the end of the ferrite rod (CH1) until the protection circuit trips. Move it just a tiny bit further away and fix in position with silicone rubber. The receive preamplifier filter inductors can be aligned using a signal near 137kHz; the tuning is very sharp.

G0MRF 300W class-D 136kHz transmitter

The following article by David Bowman, G0MRF is abridged from RadCom [19]

> No guarantees of performance are implied and no responsibility for loss or damage can be accepted. This project requires a high level of constructional experience.

With no commercial equipment available from the established 'big four' manufacturers, LF designs have relied on individuals bringing a variety of ingenious ideas to the band. This transmitter combines a number of proven techniques and, with over 300W output, allows you to get on to this exciting amateur allocation with a good signal. It is based around an amplifier using two low-cost power FETs in a high-efficiency class-D configuration. The transmitter is protected against over-current and high-VSWR conditions. The single PCB also includes forward- and reflected-power metering, output filtering and transmit /receive switching.

Circuit description

The transmit drive is generated by a pair of crystals operating as variable crystal oscillators. Crystal X1 is 8000kHz while Crystal X2 is 8274kHz (**Fig 6.8**). Each crystal is connected across a CMOS NAND gate, which functions as an oscillator. Varicap diodes are used for differential-tuning of the crystals.

The two outputs are applied to a third NAND gate which, because logic gates are non-linear, functions as a mixer. The output of IC1(c) contains several products including the difference frequency at 274kHz. A low pass filter, comprising L1, C10 and C11, removes the high-order products, leaving a sine-wave at twice the required output frequency. The filter is terminated by R7, which is part of the inverting amplifier IC2. C13 in the feedback loop adds some additional low-

Fig 6.8 (opposite): Complete circuit diagram of the G0MRF 300 watt transmitter

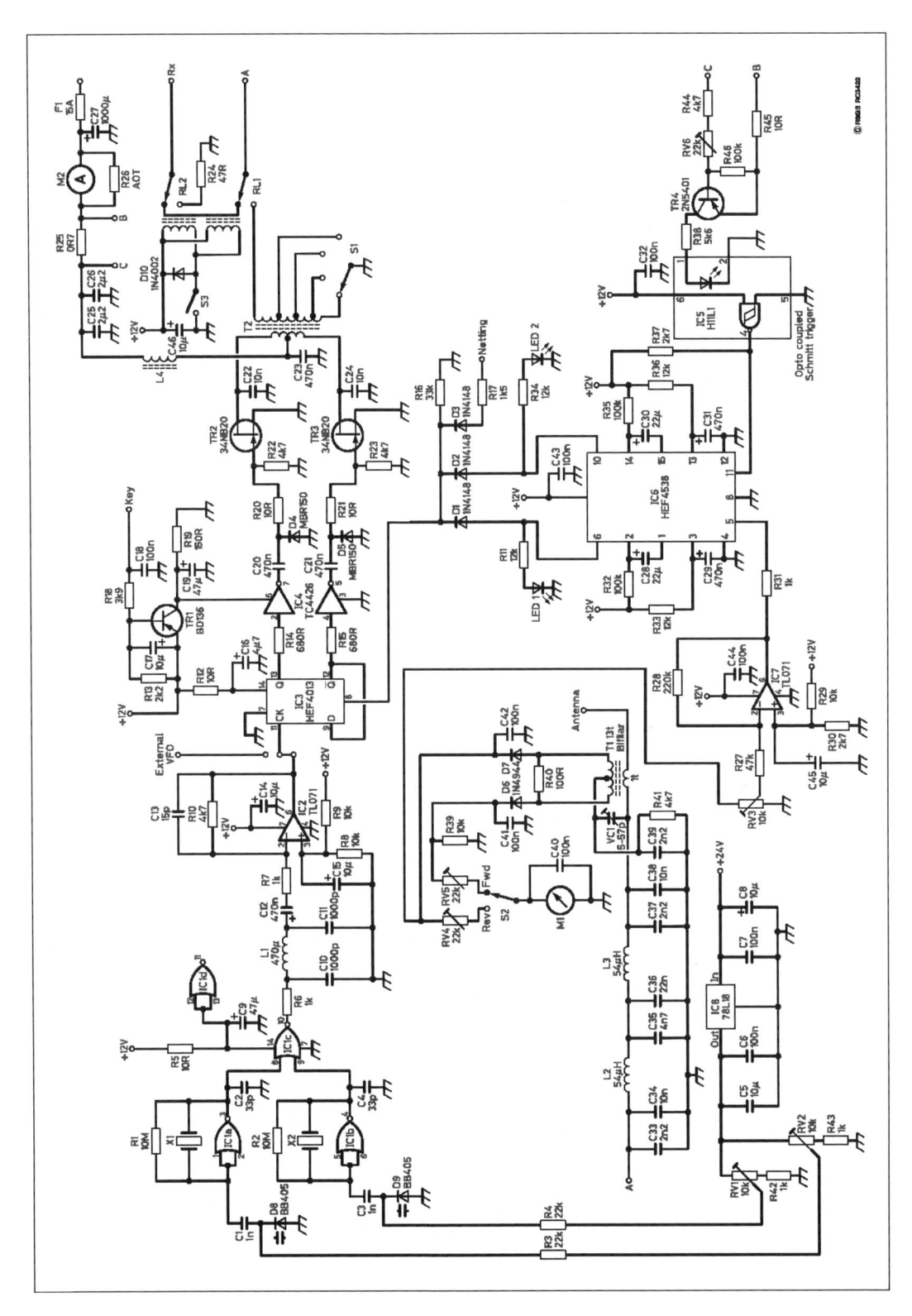

CHAPTER 6: TRANSMITTERS

pass filtering before the signal is applied to the clock input of a 4013 D-type flip-flop. A small PCB jumper provides the option of driving the transmitter from an external source.

IC3 has two functions. Firstly, it divides the input frequency by two, producing 136kHz at the (Q) and (not Q) outputs. The second function of IC3 is to act as a switch in the event of a fault condition. In normal operation the Set Direct input, pin 8, is held at 0V by R16. The circuit uses D1 - D3 as a simple discrete OR gate to provide control and protection functions. Diodes D1 and D2 feed in signals from the reflected power and over-current protection circuits, while D3 is used to provide a netting facility on receive.

Fig 6.9: Transmit-receive switching using a DPDT centre-off switch

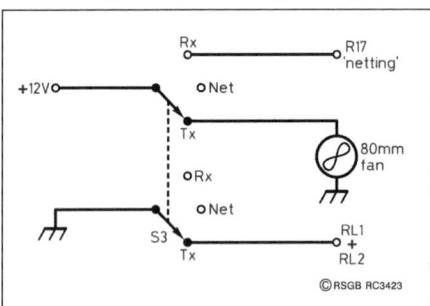

If D1 or D2 or D3 conduct, the Set Direct input will go to 12V, causing IC3 to shut down, removing the drive from the power amplifier. **Fig 6.9** shows how these functions can be controlled with a double-pole centre-off switch. One pole is used to drive the transmit / receive relays, while the other switches on a cooling fan during transmit periods.

Netting is carried out with the switch in its centre-off position. This puts the relays in the receive position, but keeps the 4013 divider active.

IC4 is a dual-inverting FET driver. It amplifies the CMOS-level signal from IC3 and is capable of driving up to 1.5A into the gates of the power FETs. A fast charge / discharge time is essential for highly efficient, and disaster-free operation of a class-D amplifier. C20 and C21 AC-couple the drive to the FETs. Schottky diodes, D4 and D5, restore the correct DC level at the gates while R20 - R23 ensure stability.

The Class-D push-pull output stage comprises FETs TR2 and TR3 and output-matching transformer, T2. Initially, I selected a large E-core transformer for T2. This ETD44 core worked well and was used in the prototype. Unfortunately, it proved both expensive and difficult to reproduce. Finally, it was replaced by a toroidal design, which was easier to construct. The drain-to-drain impedance of the FETs is matched to 50 ohms by the turns ratio of T2. A series of taps on the secondary allows the turns ratio of T2 to be changed, allowing the power delivered to the antenna to be selected via a front panel ceramic switch. The highest number of secondary turns provides the highest output power.

The DC supply is passed through an ammeter and current-sense resistor and is decoupled by C26 and C27. DC is applied to the centre tap on the primary of T2. L4 and C23 provide additional filtering.

The RF output passes from the secondary of T2 through transmit / receive relay, RL1. A second relay, RL2, has been included to terminate the receiver input when in transmit mode. These relays are rated at 12A, and have been tested at 136kHz with power levels of 1000W. Diode D10 is included to protect any semiconductors included in the external switching arrangements.

From the relays, the RF passes through a multi-element low-pass filter to the output. The LPF is essential for removing the high levels of harmonics which are present in the square-wave output from the amplifier. The T157

Front panel of the
G0MRF transmitter
[Photograph
Maurice de Silva,
G0WMD]

core used for L3 is rated to about 400W, while the polypropylene capacitors are all specified at 1kV and are capable of handling much higher power levels. The cut-off frequency of the filter is 200kHz, ensuring a very low insertion loss at 136kHz.

Reflected power protection
Forward and reflected power are detected by directional coupler, T1. A single wire passing through the centre of the toroid acts as a single-turn primary, while the secondary is a bifilar winding of 13 turns. The secondary produces outputs proportional to forward and reflected power. These AC signals are rectified by diodes D6 and D7. Preset potentiometers, RV4 and RV5, set the sensitivity. Switch S2 selects whether forward or reflected power is displayed on the meter. Resistors R29 and R30 define the reference voltage at the non-inverting input of IC7 and hence set the trip point of the protection circuit. Under normal operation, pin 6 of IC7 is at 12V. When the voltage at the wiper of RV3 exceeds the voltage at pin 3, the output at pin 6 rapidly falls from 12V to zero. This circuit was adapted from a Motorola application note [20] and is very fast-acting. It is claimed to be capable of switching off the drive in about 10 microseconds. The op-amp output is connected to the input of IC6, a 4538 dual-monostable. Once triggered, the output of the monostable changes from 0 to 12V. This voltage forward-biases D1, which causes the Set Direct function of the 4013 to shut down the device. The output of the 4538 also illuminates a front panel LED giving a visual indication of the cause of the shutdown. Having cut off the drive, the monostable maintains this condition for a period determined by R32 and C28, about 2.2s. The circuit resets automatically.

Over-current protection
Over-current protection has been implemented by utilising current-sense resistor, R25, with TR4 and opto-coupler IC5. When the current flowing through R25 causes a potential difference of 0.7V to be developed across it, the PNP transistor, TR4, will switch on. Current then flows through R38 and the diode contained within the opto-coupler.

Table 6.2: Components for the G3MRF 300W Class-D Transmitter

Resistors

R1, R2	10M
R3, R4	22k
R5, R12, R20, R21, R45	10R
R6, R7, R31, R42, R43	1k
R8, R9, R29, R39	10k
R10, R41, R44	4k7
R11, R33, R34, R36	12k
R13	2k2
R14, R15	680R
R16	33k
R17	1k5
R18	3k9
R19	150R
R22	4k7 mounted off board
R23	4k7 mounted off board
R24	47R
R25	R07, 6W, RS Comp'nts.
R26	Shunt for 15A FSD
R27	47k
R28	220k
R30, R37	2k7
R32, R35, R46	100k
R38	5k6
R40	100R
RV1, RV2	10k dual-gang pot
RV3	10k preset
RV4, RV5, RV6	22k preset

Capacitors

C1, C3	1n
C2, C4	33p
C5, C8, C14, C15	10μ
C17, C45, C46	10μ
C6, C7	100n
C9, C19	47μ
C10, C11	1n
C12, C29, C31	470n electrolytic
C13	15p
C16	4μ7 tantalum
C18, C32, C41, C42	100n ceramic 50V
C20, C21	470n polyester 63V
C22, C24	10n 50V pulse capacitors (polypropylene)
C23	470n 250V polyester
C25, C26	2μ2 100V polyester
C27	1000μ 63V - not on PCB
C28, C30	22μ tantalum 16V
C33, C38	2n2 1kV polypropylene
C34, C37	10n 1kV polypropylene
C35	4n7 1kV polypropylene
C36	22n 1kV polypropylene
C39	2n2 polystyrene 160V
C40, C43, C44	100n ceramic 50V

Semiconductors

IC1	HEF4011B
IC2	TL071CN
IC3	HEF4013B
IC4	TC4426CPA
IC5	H11L1 Opto-isolator
IC6	HEF4538B
IC7	TL071CN
IC8	78L18AZ
D1, D2, D3	1N4148
D4, D5	MBR150
D6, D7	1N4944
D8, D9	BB405 varicaps
D10	1N4002
TR1	BD136
TR2, TR3	STW34NB20 mosfet
TR4	2N5401
LED 1, LED 2, LED 3, LED 4	Two green for power: 12V & 40V. Two red ultrabright: VSWR and current trip.

Inductors

L1	470μH 7BA Toko
L2	54μH T157-2 powdered iron toroid 59 turns of 0.8mm wire
L3	54μH T157-2 powdered iron toroid 59 turns of 0.8mm wire
L4	11t 1.5mm on powdered iron toroid T94-2
T1	Pri: 13 t 0.4mm bifilar. Sec: 1t (RG58 inner) on 15mm 3C85 ferrite
T2	42mm 3C90 ferrite toroid. Pri: 10 turns 1.5mm CT. Sec: 21 turns 1mm with taps at 6, 10, 15, 18 turns.

Miscellaneous

X1	8.000 MHz crystal
X2	8.275 MHz crystal, QuartSlab. Fundamental mode, 20pF parallel load
RL1, RL2	12A relay. Single-pole change-over
VC1	5-57pF 809 series PTFE 300V trimmer, Farnell.
M1, M2	1mA FSD meter
S1	Rotary ceramic switch, single-pole 5-way, break-before-make.
S2	Forward / Reverse switch, single-pole 2-way switch.
S3	2-pole, centre-off toggle
Heat sink.	Single-sided 1.2°/W
Isolating washers	TO247
PCB	
Fan	80mm 12V, Farnell / Rapid / RS Components / Maplin, CPC, etc

CHAPTER 6: TRANSMITTERS

When the LED forward current reaches 600µA, an internal Schmitt trigger causes the output voltage to fall rapidly from 12V to zero. This triggers the other half of the dual-monostable, IC6. To preserve the speed of the overcurrent trip, there are no decoupling capacitors around TR4 or IC5. Once again, the response time of this circuit is very fast and it can reduce the output to zero in about 10-20µs. The exact value of the trip point can be adjusted over a small range by RV6.

Construction

The PCB should be assembled and tested before being fitted into an enclosure. Start by constructing the VXO and logic circuits, leaving the low-pass filter coils, transformer T2, and directional coupler T1, until last. The power FETs can be temporarily fitted for testing and then mounted permanently after testing is complete. The coils L2 and L3 in the low-pass filter are quite large and can be held in place using a little epoxy glue for extra support. When fitting inductors, ensure that the enamelled wire does not come into direct contact with the earth plane. This avoids high voltages arcing through the thin insulation and other damage due to abrasion. The primary of T1 is a single wire passing through the centre of the toroid. I used a small length of the inner conductor from RG58 coaxial cable.

The components list is shown in **Table 6.2**. The powder-coated ready-punched enclosure used in the prototype was obtained from H J Morgan Smith [21].

Testing

Testing should follow a logical procedure. Apply DC to the 12V and tuning voltage inputs. Check that the two crystals in the VXO are oscillating by looking for the 8MHz signals at pin 8 and 9 of IC1. With the oscillators running and tuning correctly, you should be able to see 274kHz at pin 6 of IC2 and 136kHz at the (Q) and (not Q) outputs of the 4013.

The range of the VXO at 136kHz will be typically 1.5kHz. While this is not sufficient to cover the entire 2.1kHz allocation, it is possible to adjust the values of C2 and C4 and select which portion of the band you wish to cover. On my prototype, I decided to cover the slow CW (QRSS) portion at the top of the band, down through the CW section to 136.3kHz.

Start by setting all the presets to mid position and VC1 to 80% mesh. Connect the output to a 50-ohm dummy load. At LF, almost any load will suffice, even wirewound resistors are a good match at 136kHz! Fit a 5A fuse temporarily to the main FET supply and switch on. Select the lowest power tap at six turns. Ground the transmit / receive pin and key the transmitter. If luck is on your side you should see 50 - 100W output. Don't be tempted to switch to high power at this point. Instead, spend some time checking the other functions at this power level.

Measure the efficiency of the amplifier and you should see a value above 70%. Values up to 86% are not uncommon. The power meter should read correctly, but if it reads backwards this can be corrected by reversing the connections on the directional coupler. With a 50Ω load, the trimmer, VC1, can be adjusted to show zero reflected power. When you are satisfied that the transmitter is operating correctly, replace the fuse with a 15A component, and test at the higher power levels. The total number of turns on the secondary of T2 has been specified as 21.

> **Warning**
> If you apply hundreds of watts at 136kHz to a high Q aerial system, the antenna voltage will very high - in excess of 20kV is not uncommon. Most insulated wire will break down and arc across to anything nearby. Keep all antenna wires well clear of everything and use good insulators. Many antennas have fallen down the moment the key was pressed, including mine. Remember that RF burns hurt and those voltages could be very dangerous!

LF TODAY

Inside the G0MRF transmitter [Photograph Maurice de Silva, G0WMD]

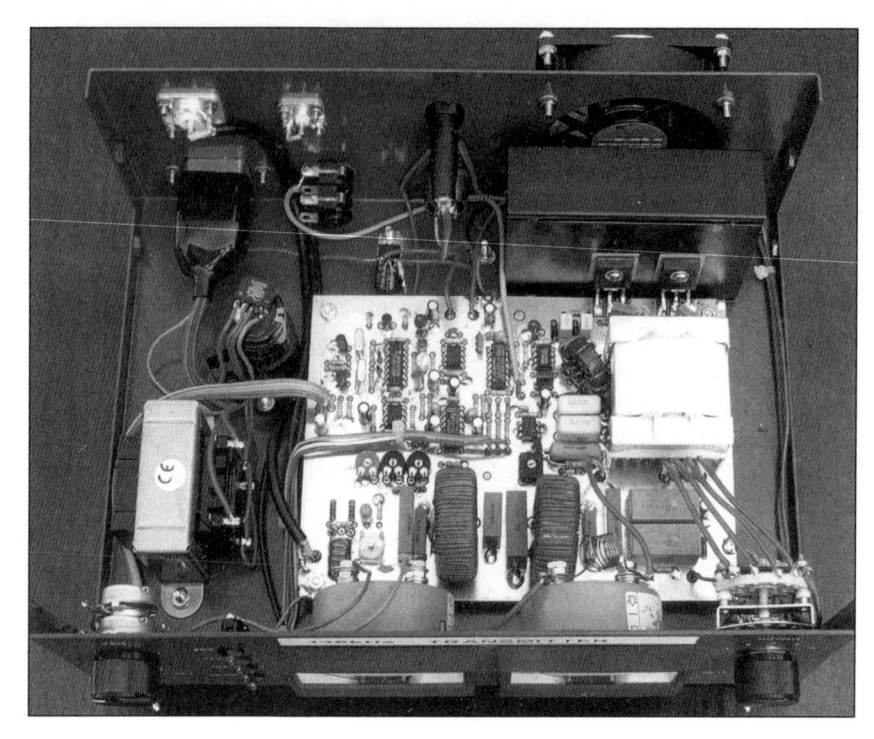

Rear panel showing the fan and connectors [Photograph Maurice de Silva, G0WMD]

In practice, you may only need to use this number of turns if you are using a supply of around 36V. If you have a higher voltage (around 45V), you will be able to achieve maximum power output using the 18-turn tap.

The final part of the setting-up procedure is to adjust the reflected power and over-current trip points. I suggest that you turn up RV3 to maximum sensitivity and see how it responds. The current trip should be set for 10-11A with a supply of 36V, but if you're using a higher voltage supply, then the trip should be arranged to cut in at 400W DC input. At 45V supply, this equals 8.9A. Preset RV6 provides a fine adjustment of the over-current trip point. Coarse adjustments can be can be made by adding 1Ω, 0.5W, resistors in parallel with R25.

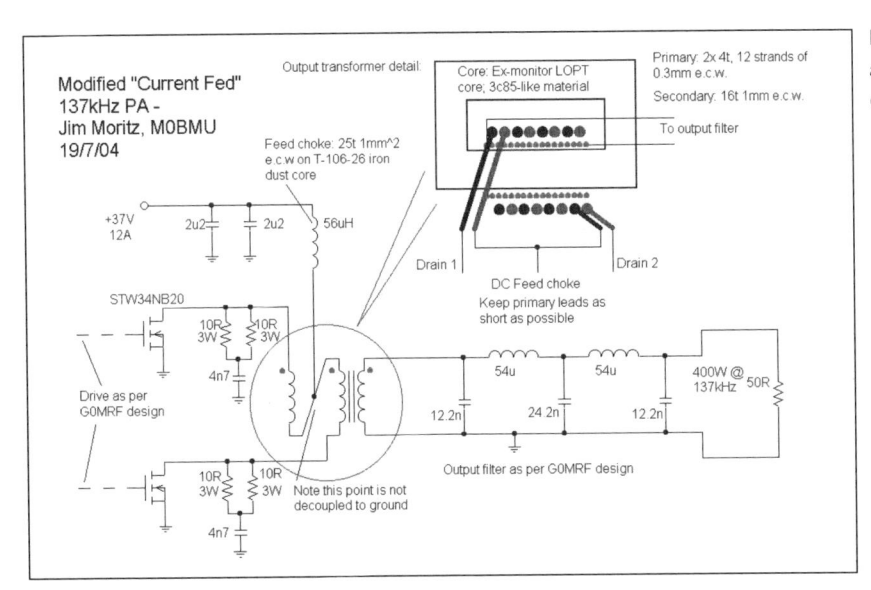

Fig 6.10: M0BMU's alternative to the G0MRF PA circuit

Alternative circuit

Following the publication of the *RadCom* article [19], Jim Moritz, M0BMU, suggested the alternative output circuit shown in **Fig 6.10**. He says:

"My unit gives 400W out with a 37V supply - for use with a 48V supply, I reckon the primary windings of the output transformer should be changed from 4 to 6 turns. The value of the choke in the DC supply is not critical - the iron dust cores can often be found in old PC SMPSUs. The most significant change to the original circuit was to remove the decoupling capacitor from the centre tap of the transformer primary and increase the inductance of the choke, converting the circuit into a 'current feed' class D configuration. This means the DC supply has a high impedance at 137kHz, which prevents high transient currents into the low pass filter capacitors when the mosfets switch. The ideal drain voltage waveform would be a half sine wave, although in practice there is still considerable ringing at the switching transitions, although much reduced. The value of the damping resistors and capacitors is a compromise; reducing R and increasing C reduces the amount of HF ringing, but increases dissipation in the resistors."

Low pass filter for 136kHz

The output from a simple LF transmitter can be high in harmonics, so it must be followed by a low pass filter.

The filter shown in **Fig 6.11** can be built as a stand-alone project so that it can be added to any experimental transmitter you may build.

The inductor is 59 turns of 0.8mm enamelled wire on a T157-2 powdered iron toroid. High voltage polypropylene capacitors with a working voltage of 400 to 1000V must be used.

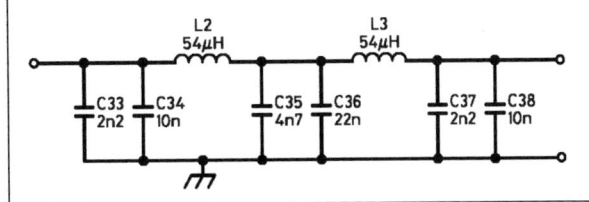

Fig 6.11: This low pass filter makes a useful stand-alone project as it can be used with any 136kHz project

Transmitters for 472kHz

Transmitters for 472kHz can be scaled-down 136kHz designs, or borrowed from HF QRP transmitter practice. Your decision on the type of transmitter to use will be based on whether you want to run relatively low power to experiment with local stations using various (particularly data) modes, or whether you want to run higher power to make CW contacts and/or work DX stations.

As with 136kHz, low-cost switching mosfets are effective as output devices, as well as devices specified as RF power amplifiers. Several amateurs successfully used simple valve transmitters on the old 500kHz band.

GW3UEP's simple MF transmitters

These MF CW transmitters were developed by GW3UEP from earlier versions used on 160m and 80m in the 1980s. When the 500kHz permit arrived in 2007, the solution was to hand and the QTX (Quick-Tx) was born! Evaluation and on-air-testing of these transmitters was thanks to collaboration with GW4HXO and EI0CF. The versions shown below are further developments to allow operation on the 472kHz band. These transmitters are in use by a number of amateur stations; further information can be found at GW3UEP's web site [22].

As with all of the transmitter designs in this book, some previous construction experience (or an experienced mentor) is highly recommended. Nevertheless the following three projects have several common factors, so it is possible to tackle them in sequence, gaining skill and knowledge as each new transmitter is built.

1 watt MF transmitter

The simple project shown in **Fig 6.12** is ideal for someone wanting to experiment on the 472kHz band for the first time. For serious operation, a VFO, more power and better output filtering (see the next two projects) is recommended.

The 7.60MHz crystal is available cheaply from the GQRP Club [23]. It is divided by 16 in IC1 to produce 475.0kHz. Other components were obtained from Rapid Electronics [24].

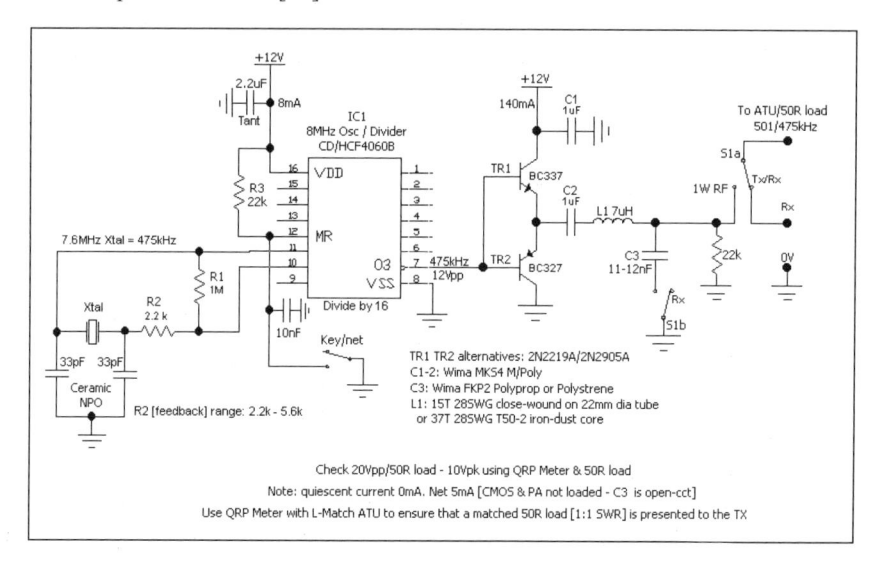

6.12: The simplest MF transmitter

CHAPTER 6: TRANSMITTERS

The complete 1W 472kHz transmitter built in a tobacco tin by GW4HXO

VFO designs

Crystal control is not recommended for anything but local QRP working, so you are advised to look at the chapter on Generating a Signal where you will find, amongst various options, GW3UEP's designs for a VFO and VXO to match these transmitters.

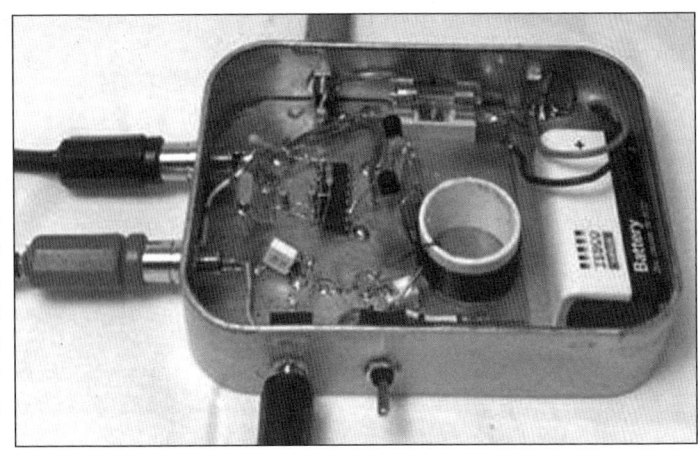

25W QTX 472kHz PA

With an output of up to 25 watts, this amplifier (**Fig 6.13**) provides a more practical power level whilst still being easily reproduceable and retaining simplicity of construction. No toroids are used; all coils being air cored. Efficiency should be greater than 80%.

The transmitter will run from a standard 12V regulated shack supply, though the voltage can be increased to a maximum of 22V if required.

Although any oscillator with sufficient drive may be used, this amplifier is designed to be compatible with GW3UEP's CMOS VFO (see the previous chapter). No more than 30cm of coax should be used to couple the oscillator to the PA.

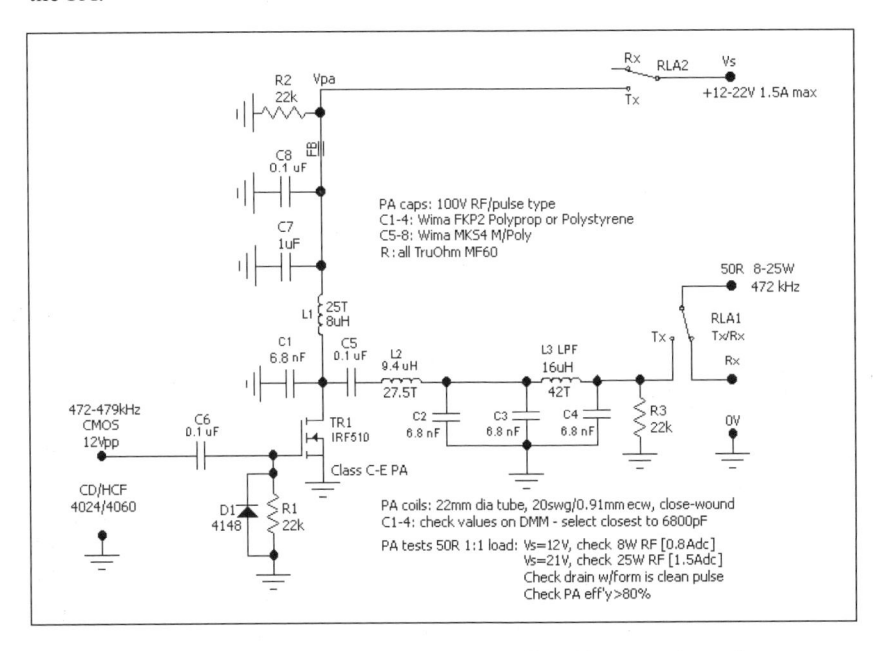

Fig 6.13: GW3UEP's QTX 25W 472kHz power amplifier. Voltages shown in square brackets are typical values

101

LF TODAY

The GW3UEP 25W 472kHz power amplifier

The amplifier can be keyed using the circuit in **Fig 6.14**, or may be left on whilst the oscillator is keyed instead.

100W 472kHz PA

For those wishing to run higher power, this amplifier produces 100W at high efficiency from a 24V regulated supply, yet is simple to build. A companion VFO can be found in the previous chapter.

Fig 6.14 shows the circuit of the amplifier. The IRF540 mosfet was chosen for operation as the power switch, TR1. Note that if other devices are used, for instance an IRF640, resistor R10 must be fitted directly on the gate pin. The PA operates in switched-mode (Class E) with drain efficiency in the 80% range.

TR3 and TR4 form a zero-biased complementary voltage follower, buffering the IC2 output stage and providing adequate source/sink current for the IRF540 gate charge (alternative devices for TR3 and TR4 are BC549/559, BC337/327, BC109/BCY71 or 2N3904/3906). The gate is AC-coupled to the buffer and DC-restored to ground, to prevent high DC current flow in TR1 should a fault in the 472kHz drive occur.

The 100W 472kHz amplifier ready for testing. This is the version without PA keying

The output circuit provides matching and LPF functions, and presents a clean sine wave into the 50 ohm load. C1/L1 forms a resonant MF tank circuit. L-match C2/L2 transforms the 50-ohm output load to a lower impedance at the drain.

The output inductors are air-cored and wound on 22mm diameter plastic 'waste pipe' available from plumbing suppliers. GW3UEP's web site [22] has a test set-up for measuring the inductance of L1-L3. Once the PA is working, the

CHAPTER 6: TRANSMITTERS

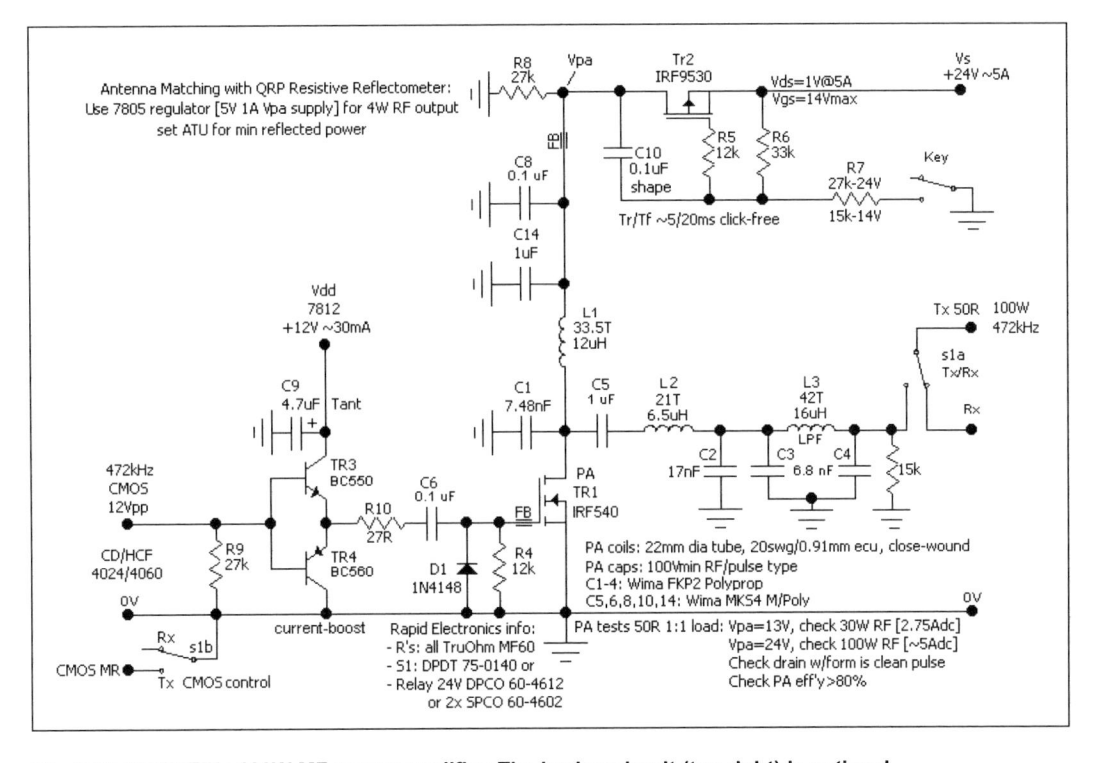

Fig 6.14: GW3UEP's 100W MF power amplifier. The keying circuit (top right) is optional

inductance can be tweaked for best results. GW3UEP uses a 'tuning wand' consisting of wooden tooth-picks fitted with a ferrite bead or dust core, and a short circuit loop of wire about 13mm diameter. Inserting the ferrite into the coil increases inductance, inserting the shorted turn reduces inductance. Once the optimum inductance is found in this way, the final coil can be constructed.

PA keying is achieved with P-channel mosfet TR2, which also shapes the keyed RF envelope and eliminates key-clicks. The key input switches TR2 gate via R6 and R7, which along with C10 also set the rise and fall times. R5 ensures stability by rolling off the frequency response of TR2, forming a LPF with its input capacitance. If keying is to be done in the oscillator stage instead, this circuitry can be omitted. TR2 is then replaced by additional transmit-receive relay contacts, which switch-off the PA drain supply on receive for key-down netting.

The maximum supply voltage for the transmitter is 25V; this allows for a voltage drop across TR2, which has VDS of 1V at 5A supply current. R6/R7 reduce the gate-source voltage of TR2 to 14V with a 25V supply. Heat sinks are required - at maximum output, TR1 dissipates 20 - 25W, TR2 5W. Ideally, a stabilised 24V PSU with current limiting set to approximately 5A should be used. Additionally, a 5A quick-blow fuse should be incorporated. An un-regulated PSU should deliver 24-25V maximum on load. Operation over the range 14-24V is recommended.

A typical setting-up procedure includes the following steps and approximate values (assumes a 24V DC supply. Measurements made using a DVM, plus oscilloscope with a 10:1 probe for the RF tests):

- Terminate the transmitter output with a 100W dummy load/power meter and observe the DC supply current.
- With no VFO input, check PA current 20mA when switched to transmit and receive [excludes relay current].
- Apply VFO input.
- With Vpa = 0, check >10Vpk-pk at 472kHz across R4.
- Apply 14V or 24V supply Vpa.
- With transmitter key-down: check 100W RF output, 5A DC supply current (Vpa = 24V) or 30W RF output and 3A (Vpa=14V).
- Check PA drain waveform is a clean pulse and that efficiency is >80%. The drain voltage waveform should be approximately 100Vpk-pk. (The drain voltage waveform for this circuit is very similar to Fig 6.16, except for the lower voltage swing.)

Attaching an antenna

It is important that the transmitter is connected to a properly matched load with an SWR of no more than 1.25:1. Details of how to tune and match an LF/MF antenna can be found in the chapter on Antennas.

G4JNT high power class E amplifier for MF

The goal of this project was originally a low cost, high power class E switch mode amplifier for 500kHz, but it should work perfectly satisfactorily on the 472kHz band. A power output of 500W running from a 50V supply was decided upon as an aiming point. Reference [15] provided the design details. A spreadsheet was developed from one by G3NYK [25] that added a series L / shunt C output matching network to raise R_{load} to 50 ohms. The series L is absorbed into either the tank circuit L or C values.

Several IRFP462 FETs were available, rated at 500V, 170W dissipation and $R_{ds(on)}$ of 0.4 ohms. The $R_{ds(on)}$ value is a bit high for the projected power level, so two devices were used in parallel.

The air-cored tank coil was wound with 2.5mm litz wire. Ceramic capacitors proved to have high losses when used in the tank circuit, and the final breadboard uses 3.3nF, 1700V metallised polypropylene capacitors built up in parallel combinations to give the required values and share the RF current.

The completed circuit in **Fig 6.15** has input circuitry consisting of a bandpass filter followed by a line receiver to get a close-to-50% duty cycle square wave from a low level input. For driving the FETs, ICL7667 gate drivers were already available. Since the ICL7667 contains two identical drivers, one was used for each parallel FET to reduce the loading due to the FET's 2000pF of input capacitance amplified by Miller feedback. The 100 microhenry DC feed choke is a large toroid, and must be capable of passing the 10A DC feed current without excessive heating.

The unit was powered initially with a 12V supply to the PA to allow tuning safely without over-stressing PA components. The PA was tuned by adding or removing 3.3nF capacitors from the parallel combinations making up the tank circuit capacitors, while viewing the switching waveform and power output (reference [15] contains a description of the tuning procedure and expected waveforms).

CHAPTER 6: TRANSMITTERS

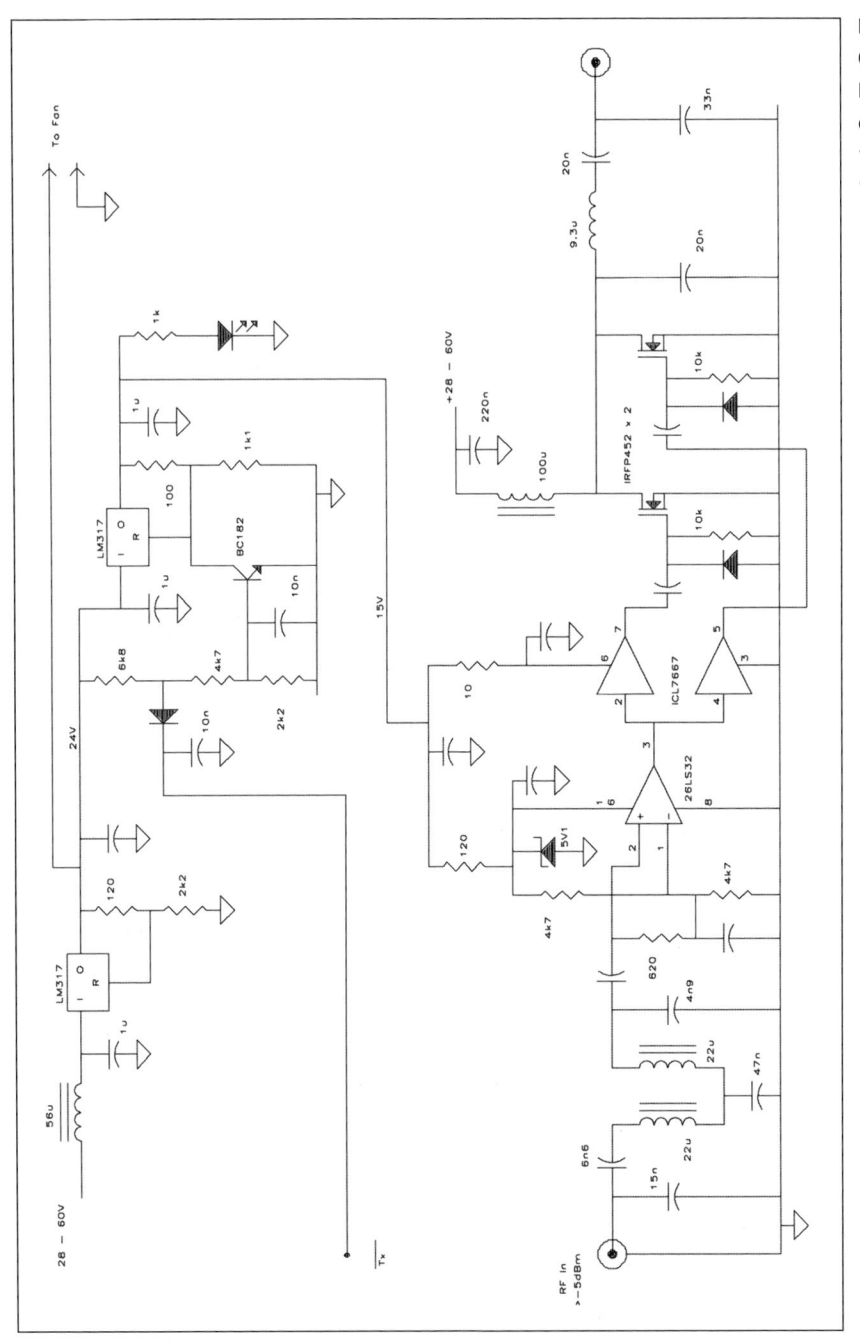

Fig 6.15: G4JNT QRO class E MF PA. All unmarked capacitors are 100nF. Diodes are 1N4145

The quantisation of the capacitance values in 3.3nF steps is a bit coarse, but combinations could be found giving close to the right waveform shape (**Fig 6.16**) and output power. With 50V, 10.1A DC input, RF output of 400W was achieved, with the heatsink running cool at this level. The peak drain voltage of 190V is also well within the FET's rating so it may be possible to safely increase the DC supply voltage to achieve the 500W target.

105

LF TODAY

Fig 6.16: Class E PA waveforms: Upper trace - Drain voltage (50V/div); Lower trace - Gate drive voltage (10V/div)

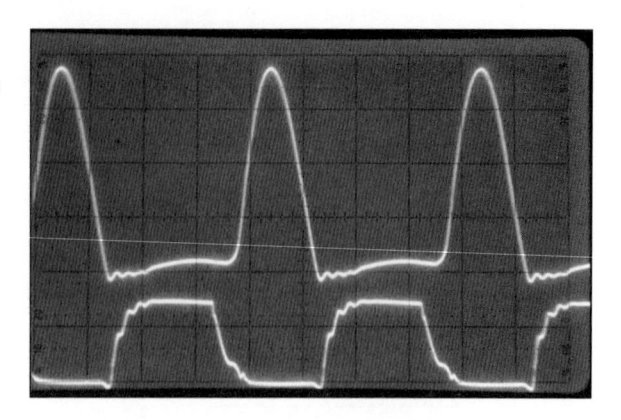

Efficiency at 79% is not the highest that can be obtained from class E PA stages - values in excess of 90% have been claimed. The major power loss in the circuit is due to the relatively high 'on' resistance of the IRFP462 FETs, as can be seen from the approximately 8V V_{ds} voltage drop visible in the waveform of **Fig 6.16**. More modern devices are available with much lower $R_{ds(on)}$.

Also the ICL7667 gate drivers may be marginal in this circuit; again more recent devices are available with higher drive capability. A more detailed article describing this project and the development process can be found at [26], and the tank circuit design spreadsheet incorporating the output matching network can also be downloaded from [27].

Low pass filter for 472kHz

The 472kHz allocation has its second harmonic in the Medium Wave AM broadcast band, so a filter with a sharp cut-off at 600kHz is desirable. The filter in **Fig 6.17**, designed by G0MRF, can also be used on receive to reduce blocking from strong AM broadcasters. The result of measurements on the prototype are shown in **Fig 6.18**.

Component values are: C1 and C4, 4400pF made from two 2200pF in parallel; C2 and C3, 10nF; L1 and L3, 20.6uH, 49 turns of 0.56mm dia on T-94-2; L2, 24.3uH, 54 turns 0.56mm on T-94-2. The capacitors are polypropylene from RS Components [13]. The maximum power level is limited by the core and wire size to about 100 watts.

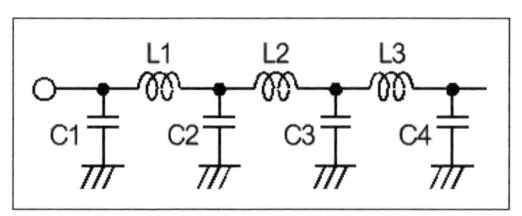

Fig 6.17: Low pass filter for 472kHz

Fig 6.18: Measured performance of G0MRF low pass filter

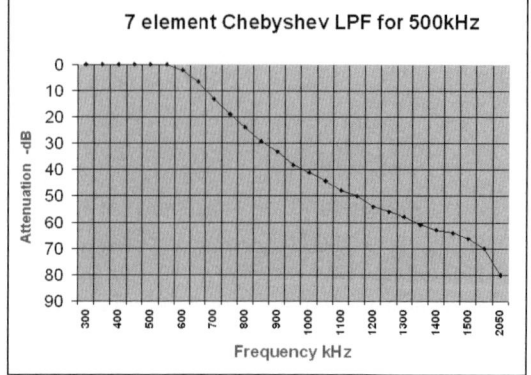

106

Modes needing linear amplification
As described above, all but very low powered transmitters for LF/MF use have tended to use Class D or Class E amplifiers. These will produce high power with relative simplicity of construction. They will also support many of the modes in common use on the 136 and 472kHz bands.

Some modes, however, require linear or quasi-linear amplification. Examples are WOLF, a highly effective DX QSO mode and PSK31 (see the chapter on Modes later). Such an amplifier may use conventional linear techniques such as seen on the HF bands, or 'EER' which is in common use in high efficiency broadcast transmitters. Both are described below in designs for both 136 and 472kHz.

G4JNT medium power linear LF/MF amplifier
This amplifier was built by G4JNT in order to have a PSK31 contact. The following are his words:

Although switching mosfets really aren't designed for linear operation, enough people have now used them with varying degrees of success to make them worth looking at. Some designs can be seen at [28]. I had a lot of spare IRF520 devices, these are 100V rated and have a quoted $R_{DS(ON)}$ of 0.2 ohms. IRF540 devices with their lower on resistance would no-doubt be more suitable but as I had so many of the former, they were just used and with a 26 - 28 volt supply, would be well within their voltage limitation. The target power output was somewhere in the 40 - 50 watt region. **Fig 6.1** shows the circuit diagram of the final amplifier.

Output transformer
Designing for maximum power and a saturation voltage of 3V; based on a 26V supply, 50 Watts output requires a push-pull load resistance (between the drains) of 232/50 * 2 = 21f¶. The design was for a nominal 50 ohms out so a turns ratio of √(50/21) = 1.54 would be needed. Looking through the junk box revealed an RM10 sized core made of F44 material, marked with an RS part number, 231-8757 (RS Components [13] still supply this item). The data sheet suggests typical operation up to 300kHz in SMPSU use, so when used with a sinusoidal waveform and lower B_{MAX} will be quite OK to 500kHz or more.

The all important core cross sectional area (A) was 95mm^2 and using the lowish value for B_{MAX} of 0.06 Tesla as a limit to avoid core losses, designing for the worst case at 135kHz the minimum number of volts per turn determined from: V_{RMS} = 4.44. F x N x A x B_{MAX} so
V/N = 4.44 x 135000Hz x 90 x 10^{-6} x 0.06 = 3.2 Volts/turn maximum (RMS volts)

With a 26V supply, Vpk-pk across the two FET drains would be 52V, so 18V RMS, and a minimum of six turns is needed. Eight turns centre-tapped were actually used, made from 4+4 parallel turns of 0.8mm wire. So for the ratio needed, a 12 turn secondary is suited. The secondary was actually made with a tapped winding of 10, 12 and 15 turns, with the taps made by twisting the wire on itself at the appropriate position / turn, and passing the double-strand across the winding to the outside world for connection. This prevents any need for soldered joints inside the transformer bobbin. The whole lot just fitted within the pot core, although it was a bit tight with all the tap positions being brought directly to the outside world.

LF TODAY

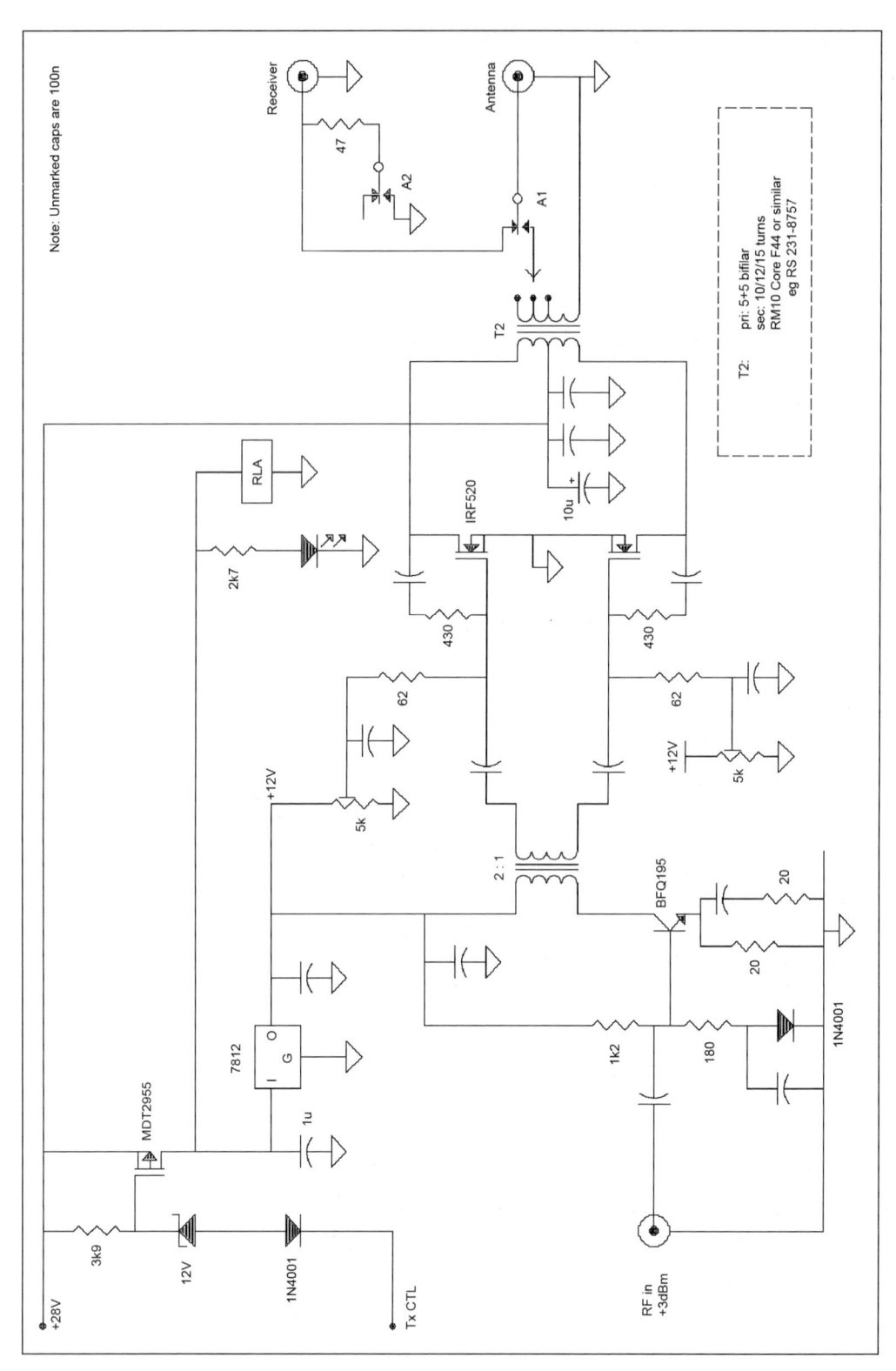

Fig 6.19: G4JNT's medium power linear amplifier for 136kHz and 472kHz

Driver stage

To keep things simple, the mosfet gates are damped with 62 ohm resistors, which also serve as the bias inject - bias voltage being set separately by individual presets for each device supplied from a stabilised 12V source. Differential impedance is therefore a maximum of 120 ohms resistive, and with Miller feedback from drain to gate, we are probably looking at appreciably lower value of R_{in}.

Each device was set to run with around 100mA bias. The devices can be driven to saturation with 250mW from a 50 ohm source applied differentially to the gates. 150mW was sufficient for just-noticeable nonlinearity in the output waveform. To compensate the higher G_m, and the fact the FETs were non-optimal, switching devices compared with proper RF ones, a simple feedback network consisting of a 430 ohm 0.5W resistor and DC blocking capacitor was added between drain and gate of each device. This modest negative feedback lowered the gain such that around 300mW was now needed for full drive, and the total input resistance reduced to around 40 ohms.

As a driver I wanted to use an SMT device with a PCB pad as its heatsink. The BFQ19S has a ridiculously high Ft for use at this frequency (over 5GHz), but was to hand, is readily available cheap, and has the right sort of power rating, Pmax = 1 watt. Running from the 12V stabilised rail, a load resistance of 160 ohms will allow 300mW output maximum with a quiescent current of 60mA. All comfortably within its ratings. So an intermediate transformer of 2:1 is now needed to transform this to the 40 ohms or so of the device inputs. A small 12mm toroid, of (probably) 3C85 or F44 type material was wound with five quadrifilar turns of 0.2mm wire. Two of the four strands were series connected for the primary with the remaining two paralleled for a thicker secondary, giving a 2:1 isolated transmission line transformer.

I needed to obtain full power with less than +3dBm drive, so the driver had to have about 24dB gain (16 times voltage). With R_{load} for the BFQ19S of 160 ohms an emitter degeneration resistor of about 160/16 = 10 ohms was required. This was made up of a 20 ohm unit setting the DC bias, and another 20 ohms in

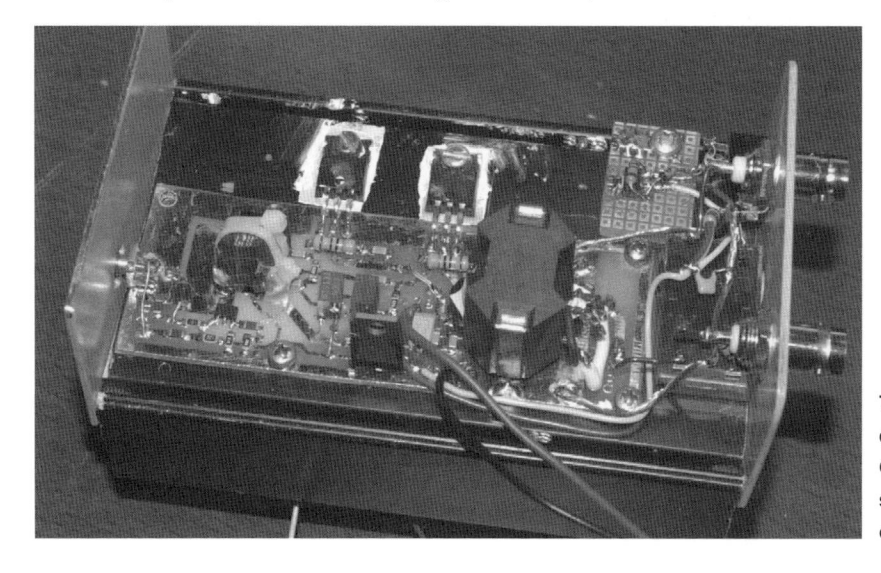

The finished design of the G4JNT linear using surface mount components

parallel, decoupled at AC. The second one can be varied to adjust the overall gain. Base bias resistors were chosen to give the 60mA quiescent, and present a load to the input of about 50 ohms. The IN4001 compensates bias with temperature.

Control and switching
The PA is enabled and disabled from a ground-to-Tx line. A P-Channel mosfet controls the input to the 12V regulator, so all bias and driver supplies are removed in standby, resulting in zero power consumption. A small 24V fan was also wired into circuit to allow a smaller heatsink to be used than might otherwise be reasonable.

A double pole relay was used for antenna changeover. One set of contacts are used conventionally, and the other set used to switch a 47 ohm resistor across the Rx port when transmitting. This lowers RF leakage and ensured the receiver input is correctly terminated. An isolation of 70dB was measured at 475kHz.

Results
Maximum power output before device saturation sets in is around 40 - 50 watts with a 26V supply (using the 15 turn secondary tap into 50 ohms for a device R_{load} of 14 ohms. Lower output at 35 - 40 Watts at slightly reduced current (higher efficiency) was possible with the 12 turn tap position. Much above this the waveform began to flatten. The devices saturate with about 3 - 4V across them, so efficiency is lower than could be possible.

Using IRF540 devices would give lower V_{DS} for increased efficiency At 40W into a resistive load, the output sinewave looked perfect on a scope, and on a spectrum analyser harmonics showed to be in the -30dB region. Given the Q of the antenna system - no Low Pass filter was going to be needed provided the amplifier wasn't driven into saturation.

Outline of a 136/472kHz EER transverter

The following was abridged from an article [29] by Jim Moritz, M0BMU.

Envelope elimination and restoration (Kahn technique, EER) is a method of generating amplitude- and phase-modulated signals whilst using a high efficiency, non-linear PA, without introducing excessive distortion. This allows simple, conventional LF/MF amplifiers to be used for modes where a linear amplifier would normally be required.

The signal to be amplified is divided into a carrier phase channel and an amplitude envelope channel. The constant-amplitude, phase-modulated carrier frequency is applied to the PA input, which can be a class D or E switching mode type to achieve high efficiency. The amplitude modulation signal is used to modulate the DC supply voltage to the PA. The output from the PA retains the phase modulation in the carrier signal, but now also varies in amplitude proportional to the envelope modulation signal.

Since any type of signal is effectively a carrier frequency with a combination of amplitude and phase modulation, the EER technique can in principle be used with any form of modulation. If the high efficiency amplifier is used in combination with a switch-mode modulator, very high efficiencies are possible, which makes the technique popular for high-power AM/SSB/digital broadcast transmitters.

CHAPTER 6: TRANSMITTERS

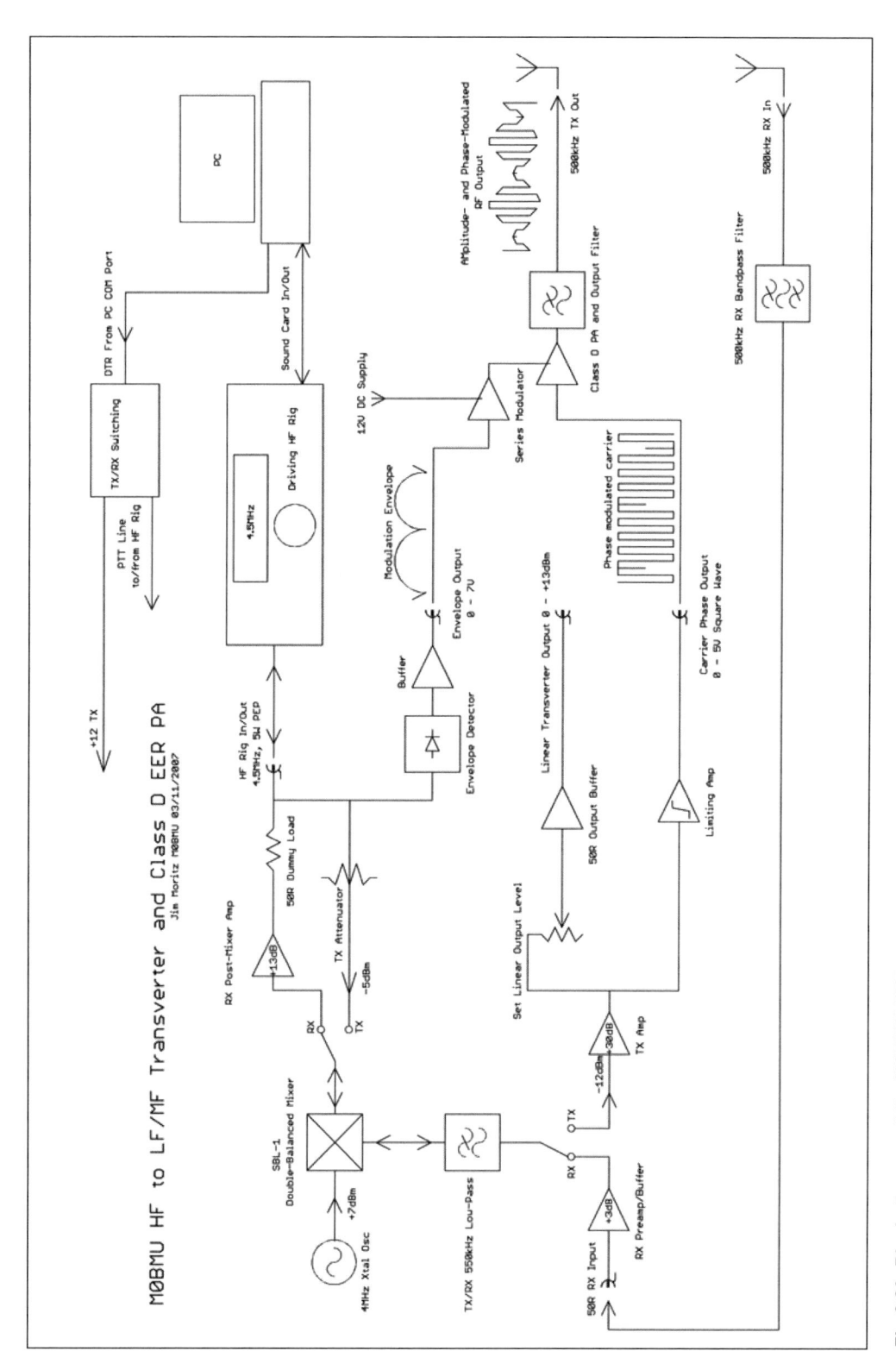

Fig 6.20: Block diagram of the M0BMU EER transverter

LF TODAY

For these relatively high power, wide-band broadcast transmitters, complicated techniques are required to ensure alignment between amplitude and phase channels. But for LF/MF amateur applications at moderate power levels and using narrow-band modulation, satisfactory signal quality can be achieved quite easily.

A design by M0BMU (**Fig 6.20**) uses a simple linear modulator which dissipates a significant amount of the DC input power. For types of modulation with a reasonably high crest factor (ie ratio between average power and PEP) the losses in the modulator are quite low. For example the efficiency calculated for some common types of modulation, assuming an idealised system with a 100% efficient PA and a series modulator that delivers the full supply voltage to the PA at modulation peaks gives the results shown in **Table 10.3**.

Table 10:3: Theoretical efficiencies achievable using EER

Mode types	Efficiency
A1A (CW), FSK (no envelope modulation):	100%
BPSK with envelope shaping (eg PSK31):	89%
Two tone 'IFK' modulation (eg Throb):	78%

In practice, there are additional losses in both PA and modulator, so in this design actual efficiency is perhaps 20% less than these figures, but this is still better than a class AB linear PA transmitting the same modes.

A block diagram of M0BMU's transverter system is shown in Fig 6.20. The two main objectives of this design were to provide a convenient way of transmitting and receiving digital mode signals on the 136kHz and 500kHz amateur bands, and to act as a test bed for transmission of different modes using the EER technique for power amplification.

The transverter signal source is an HF SSB transceiver, running 5W ERP, with audio input and output from a PC sound card. This enables the use of a wide range of 'sound card mode' software, together with the convenient VFO and filter facilities of the HF rig.

The basic transverter mixes this signal with a 4MHz local oscillator to obtain a low-level output in the range 20kHz - 550kHz. For reception, the signal path is reversed to convert the LF/MF input signal up to the 4.0 - 4.5MHz range. Transmit output at 200W PEP is achieved using a class D PA and modulator using EER, although the transverter also has a low-level output for use with a conventional linear PA (see above). The transverter IF input/output circuit is wide-band, so other input frequency ranges could be used just by changing the crystal frequency.

In the transverter, the carrier phase and amplitude signals are separated from the modulated signal from the HF transceiver. The carrier phase signal is obtained by feeding the down-converted signal into a limiting amplifier. The limiter output is a logic-level square wave at constant amplitude. The envelope modulation signal is obtained by rectifying the HF signal with a diode envelope detector, buffered by an op-amp follower.

The modulator and PAs are designed for use with 13.8V DC. The objective was to produce a rig for possible future portable battery operation, although it is also convenient for use with the high current 13.8V PSUs present in many shacks. The PA circuits draw about 19A at full output.

Separate narrow-band power amplifier circuits were used for 136kHz and 500kHz. Since the mosfet and driver components are cheap, and band switching would be quite awkward due to the high currents and low impedances involved, it was felt better to have separate PAs for each band rather than trying to produce a dual band design.

This transverter has so far been used successfully with a wide range of digital modes, including RTTY, PSK31, 'Olivia' and weak-signal modes such as WSPR and JT65. Also, it has been used for conventional CW and 'visual' modes such as QRSS and DF6NM's Chirped Hellschreiber using DL4YHF's *Spectrum Lab* software. All that is necessary to change modes is to load the appropriate software into the PC and 'follow the instructions'. For more on this, see the Modes chapter.

Full schematics and description of the transverter, modulator and power amplifiers can be found at [29].

Power supplies

Most amateur radio shacks have a high current 12 volt stabilised supply, so it may seem obvious to use that for a low frequency transmitter. Indeed some of the transmitters shown above are designed for that voltage.

Higher voltages are often used, however, with power FETs because it keeps the current down, and hence reduces the need for very thick wiring and very short leads.

A simple 100V supply is shown as part of the G3YXM kilowatt transmitter (Fig 6.6), together with a low-current stabilised 12V supply for the VFO. The high voltage line is unstabilised as the PA is not required to be linear - it is either on or off. Any stabilisation would in any case be complex and would generate a great deal of heat.

It is possible to find surplus high voltage, high current, stabilised supplies, for instance those formerly used in telephone exchanges, at rallies. Chargers for computer back up supplies can also be useful.

Take great care when using high voltages and/or high currents. Avoid working on equipment when it is powered up. If you must do so, remove metal objects such as rings or watches, and keep one hand in your pocket to avoid an electric shock across your heart. Make sure someone knows where you are, and how to turn off any power.

References

[1] Juma kits: *http://www.nikkemedia.fi/juma/*
[2] TX-2200 suppliers: *http://www.wsplc.com*, & *http://www.radioworld.co.uk*
[3] TX-2200 spec: *http://www.thamway.co.jp/ham/ham_02_e.html*
[4] Ropex Tx review: *http://wireless.org.uk/ropex.htm*
[5] The Decca system: *http://www.radarpages.co.uk/mob/navaids/decca/decca1.htm*
[6] *http://homepage.ntlworld.com/mike.dennison/index/lf/decca/transmitter.htm* - Decca transmitter circuits
[7] BK Electronics, Units 1, 3 and 5 Comet Way, Southend-On-Sea Essex SS2 6TR. Tel: 01702-527572. *http://www.bkelec.com/*
[8] *http://homepage.ntlworld.com/mike.dennison/index/lf/station/bk_electronics_amp.htm* - BK audio amplifiers as LF transmitters
[9] *http://www.hafler.com/techsupport/pdf/P3000_datasheet.pdf* - Hafler audio amplifiers

[10] Linear amp in use: *http://www.w1tag.com/XESTX.htm*
[11] *http://www.radio-electronics.com/info/rf-technology-design/coupler-combiner-splitter/wilkinson-splitter-combiner-divider.php* - Wilkinson combiner tutorial
[12] G4JNT's 600W PA: *http://www.wireless.org.uk/jnt.htm*
[13] *http://uk.rs-online.com/web/* - RS Components on-line catalogue
[14] Farnell web pages & on-line catalogue: *http://www.farnell.com*
[15] *http://www.cs.berkeley.edu/~culler/AIIT/papers/radio/Sokal%20AACD5-poweramps.pdf* - description of class E PA operation, including design formulas and tuning procedure.
[16] *http://www.g4jnt.com/DownLoad/classe_match.xls* - design spreadsheet for class E PA, including impedance matching section.
[17] *http://www.g4jnt.com/QRO_500kHz_PA_Breadboard.pdf* - G4JNT 500kHz class E PA
[18] Transmitters for 136kHz: *http://www.wireless.org.uk/136rig.htm*
[19] A Class-D Transmitter for 136kHz, by David Bowman, G0MRF, *RadCom*, Jan/Feb 2003, RSGB
[20] 'AR510: VSWR Protection of Solid State RF Power Amplifiers', by H O Granberg, *RF Design*, Feb 1991.
[21] H J Morgan Smith, sheet metal engineers. Tel: 01293 452 421.
[22] GW3UEP's projects: *http://www.gw3uep.ukfsn.org/index.htm*
[23] Source of crystals: *http://www.gqrp.com/sales.htm*
[24] Rapid Electronics: *http://www.rapidonline.com/*
[25] G3NYK's spreadsheet was not available online at the time of publication, however a Class-E design spreadsheet by WA0ITP can be found at: *http://www.wa0itp.com/class%20e%20design.html*
[26] *http://www.g4jnt.com/QRO_500kHz_PA_Breadboard.pdf* - G4JNT 500kHz class E PA
[27] *http://www.g4jnt.com/DownLoad/classe_match.xls* - design spreadsheet for class E PA, including impedance matching section.
[28] *http://www.g0mrf.com/lf.htm*
[29] *Radio Communication Handbook, 11th ed*, RSGB.

7

Measurement and calculation

In this chapter:

- Test gear for setting up the station
- Dummy loads
- Measuring a working station
- Estimating radiated power
- Field strength measurement
- Scopematch tuning aid
- 472kHz antenna tuning meter
- Frequency stability/calibration

TO GET THE BEST OUT OF your low frequency station, it will be necessary to use a combination of readily available and home constructed test gear. The basic requirements are to be able to measure the transmitted frequency and the radiated power.

It is no longer necessary to demonstrate that you know how to measure radiated power in order to obtain a permit for MF operation. However, the licence conditions for the 136kHz and 472kHz bands require you to stay within the power limits. At the time of writing, the UK limits are 1W ERP on 136kHz and 5W EIRP on 472kHz (see below for the difference between ERP and EIRP).

ERP can be calculated either from measurements of the antenna current and dimensions, or, more directly, by measuring the field strength produced by the station. Both these approaches are described in this chapter.

Additionally, it is useful to have a convenient means of checking that the antenna system has been tuned to resonance. For anyone building their own transmitter, it may be necessary to obtain, or borrow, a good range of test gear, including an oscilloscope. In this chapter you will find the calculations necessary to design, test and ultimately improve your low frequency station.

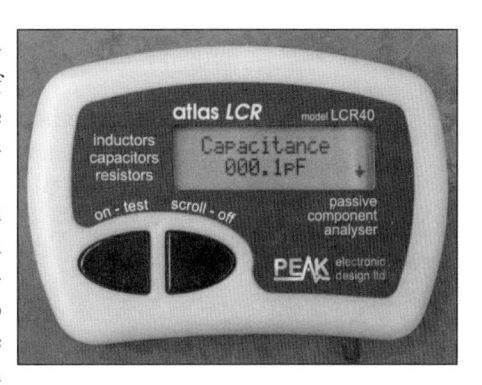

This small box measures resistance, capacitance and inductance over a wide range of values

Test gear for setting up the station

The most obvious piece of test gear for the constructor is a multimeter. Preferably you should have both a digital meter for accuracy and an analogue meter for reading the changing results when making adjustments.

An oscilloscope is a boon for development or testing a homebrew transmitter, since examining the waveforms and measuring signal levels at various points in the circuit allows correct circuit operation to be verified.

Fig 7.1: A simple RF probe is a useful tool for constructors

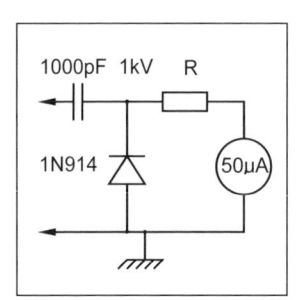

For experimenting with the construction of large loading coils, an inductance meter is useful. These are available for a few tens of pounds and cover from one or two millihenries to several henries. Check that the range is suitable for the coils you want to build.

The simplest item to build is an RF probe (**Fig 7.1**) which has many uses for anyone building a transmitter or even an oscillator.

A frequency counter can be helpful to check that the transmitter's stages are working at the correct frequency. A great deal can be done simply by using the station receiver to check on frequencies, harmonics, sub-harmonics and distortion.

Some low frequency operators have managed to buy a second hand Selective Level Meter (SLM) originally used in analogue telecommunications systems. As well as being useful as receivers, SLMs provide accurate signal level measurements over a wide dynamic range. This makes them useful for measurement of spurious and harmonic output levels of transmitters and numerous other types of measurement, especially field strength measurements when used in conjunction with a calibrated antenna..

For setting up the resonating and matching of low frequency antennas, G3LDO recommends using the Array Solutions AIM4170 Antenna Analyser [1].

Dummy loads

Conventional dummy loads intended for the HF and higher bands will of course work perfectly well at LF and MF. However, higher powers are often required, particularly for 136kHz. Fortunately, many types of relatively cheap, high power resistors will give good results at low frequencies.

Wirewound resistors with values from a few tens to a few hundred ohms usually have manageable inductance at low frequencies. For example, a 50 ohm load made up of two 100Ω, 150W (Arcol HS150 metal clad wirewound) resistors in parallel was found to have an impedance of (50.5 + j6.2 Ω) at 137kHz. This amounts to an SWR of 1.13:1, which would be adequate for many applications. Connecting two 1.5nF, 1kV polypropylene capacitors in parallel with the resistors to 'tune out' the inductive component of the impedance resulted in a SWR of 1.02:1.

More recently, 'power metal film' resistors have become available at reasonable cost in ratings of tens to hundreds of watts (eg Vishay RCH50 series rated at 50W, available from RS components [2]). These have very low inductance and give good performance as dummy loads into the HF range and above. Like the

A selective level meter is designed for accurate measurement of LF/MF signals on a telecom network

metal-clad wirewound resistors, these are designed to be bolted to a heatsink, and suitable series-parallel combinations can be used to build a high power load.

Several amateurs have pressed various types of heating element into service as high power 136kHz loads, including such things as toasters and electric fan heaters. These often have resistance in the vicinity of 50 ohms, with elements of wire or strip in a zig-zag shape that has a reasonably low inductance. It is advisable to run such a load at considerably less than its mains rated power, since the resistance rises considerably when at its normal operating temperature.

Measuring a working station

RF current

The most important piece of test equipment when transmitting at low frequencies is an antenna current meter - actually this should be called 'antenna system' meter as it is sometimes convenient to measure the current in the earth wire, not the antenna. The two most common ways of measuring current are a thermocouple meter and a transformer-coupled meter. Do not try to use the current range of an ordinary multi-meter to measure RF.

The range required depends on your output power and the type and size of your antenna. A Marconi antenna is likely to be fed with 0.5 to 4A on 136kHz, and rather less than this at 472kHz; a loop antenna may be several tens of amps. It is useful to have a really low current meter as well for low power experiments or for tuning your antenna.

RF ammeters using the heating of metals to produce a voltage - thermocouples - are available from time to time on the surplus market. Although they are convenient, they have square-law scales which can be difficult to read at the low end and can easily burn out (permanently) if overloaded. If you can get hold of one, use it to check your transformer-type meter (see below), then put it away safely.

Fig 7.2 shows a much more useful RF meter that can be easily constructed and can be switched to cover several current ranges. Although the description that follows is for a current meter with an FSD of 1A the information given will enable you to construct a meter with any current range.

A small transformer wound on a high permeability (5000 or above) ferrite core can be used to precisely sample the current flowing in a conductor. The sampled current flowing from the secondary winding of the transformer is equal to the primary current multiplied by the transformer turns ratio:

(primary turns) / (secondary turns)

Rectifying the sample and applying the output to a moving coil meter provides a predictable and reliable method of current indication which will tolerate overloads, is linear scaled and will respond quickly.

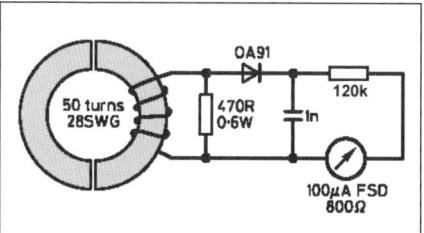

Most secondhand thermocouple RF meters are stand-alone, but this Russian one is in a bakelite case

Fig 7.2: This RF current meter is linear and versatile. The ferrite ring can be split to enable it to be clamped onto a wire without disconnecting it

With a single turn on the primary side and 50 turns on the secondary we can expect precisely 20mA to circulate in the secondary for each amp of primary current (1A / (50) = 0.02. In dealing with toroid ring cores a 'turn' simply means a pass through the central hole - it does not need to be complete.

The secondary load resistance must be small compared to the secondary winding impedance in order to get an accurate current ratio. With typical high-permeability ferrite toroids, a 50 turn winding will have an impedance of the order of kilohms. A load of 470 ohms is suitable. Due to the 50:1 transformer ratio, the transformed load in the primary will be (470 /502), or 0.18 ohms, which is low enough to have minimal insertion loss in a typical vertical antenna system. A larger turns ratio will be required for loop antenna current measurements.

With 20mA passing through 470 ohms we will have 9.4 volts RMS available for rectification, corresponding to 13.2 volts peak. We can expect to lose approximately 0.5 volts at the detector diode leaving a DC voltage of 12.7 to drive the moving coil meter. To obtain full scale deflection of 100μA with 12.7 volts requires a total resistance of 127 kilohms. The meter itself will contribute about 800 ohms so a practical 120k resistor will fit the bill with negligible error. Finally. the effective resistance of the detector circuit will be much greater than the 470 ohms, so there will be negligible error due to loading here.

A useful variant is to use a split toroid so that the meter can be clamped onto an existing wire, such as an earthing cable, rather than having the wire passed through it. There's more on current meters, including calibration and making a clip-on meter at G3SEK's web site [3].

SWR

SWR bridges, as widely used at HF or VHF, can also be used on lower frequencies. The SWR bridge indicates the degree of mismatch between the impedance at the antenna feeder and the design impedance level of the SWR meter (normally 50 ohms), and so is useful for adjusting and monitoring the antenna tuning. Commercial SWR meters intended for the HF bands may be usable, but usually have reduced sensitivity at lower frequencies. Check for correct operation with a dummy load before using with a real antenna. An examples of an SWR bridge designed specifically for LF/MF use is included in one of the designs in the Transmitters chapter. Some SWR bridges provide a power measurement facility but, at 136kHz and 472kHz, antenna current measurement is much more useful for power determination, as will be discussed in the next section.

Estimating radiated power by calculation

Licensed amateurs will be familiar with calculating the DC input power to a transmitter, which is the DC voltage applied to the PA, multiplied by the DC current drawn by the PA:

$$P_{DC} = V_{DC} \times I_{DC}$$

Also familiar is calculation of RF output power into a load resistance R if the RMS RF voltage or current in the load is known:

$P_{RF} = I^2R$, or $P_{RF} = V^2/R$,

Which allows calculation of transmitter power delivered to a resistive load, eg a dummy load, or an antenna tuned to resonance.

However, the licence conditions for the low bands allocations specify a power limit in terms of radiated power; ERP on 136kHz or EIRP on 472kHz. To calculate this power, we need to know how much power is actually radiated by the antenna, and what directivity (ie directional gain) the antenna has. At HF and above, we are used to the idea that the radiated power will be close to the transmitter output, because antennas in this frequency range have quite high efficiency. However, the efficiency of amateur antennas in the LF/MF range varies greatly, and is extremely small - a fraction of one per cent - so the radiated power will be much less than the amount of RF power fed into the antenna. So we must take into account the efficiency of the antenna to calculate radiated power.

The difference between ERP and EIRP
EIRP and ERP are defined as the product of the power supplied to the antenna and the antenna gain. The primary difference between them is that for ERP (effective radiated power), the antenna gain is expressed relative to an ideal half-wave dipole antenna whereas with EIRP (effective isotropic radiated power), the antenna gain is expressed relative to a theoretical ideal omni-directional 'isotropic' antenna.

ERP: For 'electrically small' Marconi (vertical) or loop antennas, (ie antennas whose dimensions are a small fraction of a wavelength) the directivity can be assumed to be 1.8 (2.62dB) compared to a reference dipole. It sounds paradoxical that a small, inefficient antenna can have 'gain' over a dipole, but directivity is only a way of including the difference in directional patterns of the antennas in the calculation, and so takes no account of efficiency.

EIRP: A dipole is said to have a directional 'gain' of 1.4 (2.15dB) over an isotropic antenna. Therefore, a low frequency antenna has a directional gain over an isotropic source of 1.8 x 1.4 (2.62 + 2.15dB), a total of 2.52 times (4.77dB).

So in quantitative terms, for the same RF power and antenna, radiated power measured as EIRP is 1.4 times (2.15dB) greater than that measured as ERP.

Of course, practical low frequency antennas have negative gains over full-sized antennas, but directional gains still need to be factored into any calculation of radiated power.

Calculating antenna efficiency
The efficiency of an antenna as a percentage is:

100% x R_{rad} / ($R_{rad}+R_{loss}$)

where R_{rad} is the radiation resistance, and R_{loss} is the loss resistance of the antenna.

Calculating radiation resistance: 1 -Marconi type antennas
The radiation resistance represents the proportion of the power fed into the antenna that is converted into a radiated signal. R_{rad} can be calculated for a

Marconi (vertical) antenna from the effective height H_{eff}, and the wavelength λ, using the well-known formula:

$$R_{rad} = 160\pi^2 H_{eff}^2 / \lambda^2$$

where the wavelength $\lambda = (3 \times 10^8)/f$, where λ is in metres and f hertz.

Effective height was described in the chapter on transmitting antennas, and it can be calculated from the antenna dimensions. For simple antenna shapes, such as the L and T configurations, the following formula can be used:

$$H_{eff} = H_{actual} \times (2C_h + C_v) / 2(C_h + C_v)$$

where H_{actual} is the actual height of the top loading wires above ground, C_h is the capacitance to ground of the top loading wires, and C_v is the capacitance to ground of the vertical downlead. A reasonably accurate estimate of C_h is 5pF per metre of wire, and C_v 6pF/m. See the Antennas chapter for a discussion of the effect of multiple wires.

Example:
A Marconi vertical antenna has a top loading wire 40m long at an actual height above ground of 13m. $C_h = 40 \times 5pF = 200pF$, $C_v = 13 \times 6pF = 78pF$, $H_{eff} = 11.2m$. At 137kHz, $\lambda = 2190m$, $R_{rad} = 0.041$ ohms.

H_{eff} will be between 50% and 100% of the actual height of the antenna, depending on the length of the top loading wires. When the top loading wires are sloping, for example in the umbrella configuration, the above formula can be used by making H_{actual} the average height of the sloping wires.

Calculating radiation resistance: 2 - loop antennas

For a loop antenna, the radiation resistance depends on the enclosed area A of the loop and the wavelength:

$$R_{rad} = 2 \times 320\pi^4 A^2 / \lambda^4$$

(The factor of 2 in the formula is due to the presence of the ground plane under the antenna, which increases R_{rad}.) The shape of the loop does not affect R_{rad}.

Example:
G3YMC's loop in Chapter 3 has enclosed area of 100m². At 137kHz, $\lambda = 2190m$, $R_{rad} = 27\mu\Omega$ (micro-ohms).

Calculating loss resistance

The loss resistance R_{loss} (**Fig 7.3**) represents the proportion of the power fed to the antenna that is lost, being dissipated as heat in the antenna wire, tuner, ground, and lossy objects near the antenna. In amateur antennas, R_{loss} is invariably much larger than R_{rad}.

The loss resistance can be found by measuring the RF resistance at the antenna feed point when the antenna is tuned to resonance (strictly, this will be $R_{rad} + R_{loss}$, but in practice R_{loss} is so much larger than R_{rad} that the antenna resistance can be taken as equal to R_{loss}). R_{loss} could also be found by measuring the transmitter power output and the antenna current I when the antenna is at resonance:

$R_{loss} = P_{RF} / I^2$

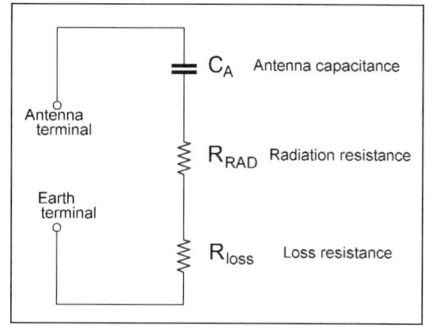

Fig 7.3: Equivalent circuit of vertical antenna

Calculating radiated power
Once R_{rad} is calculated, and R_{loss} determined, the efficiency can be calculated, and the radiated power (not ERP - see below) is then:

Tx power x Efficiency/100%

An easier and more accurate approach to finding the radiated power, which avoids the need for RF impedance and transmitter power measurements, is to measure the antenna current. Radiated power P_{rad} is then just:

$P_{rad} = I^2 R_{rad}$

To get ERP (for the 136kHz band), and EIRP (for the 472kHz band) we multiply the radiated power by the antenna directivity as described above:

$P_{ERP} = P_{rad}$ x 1.8 = 1.8 x I^2 x R_{rad}
$P_{EIRP} = P_{rad}$ x 2.52 = 2.52 x I^2 x R_{rad}

The formula for P_{ERP} (or P_{EIRP}) and for R_{rad} can be combined to give ERP or EIRP directly for the Marconi or loop antennas as shown below.

Calculating ERP and EIRP for a Marconi (vertical) antenna

$P_{ERP} = 1.8$ x 160 π^2 I^2 H_{eff}^2 / λ^2
$P_{EIRP} = 2.52$ x 160 π^2 I^2 H_{eff}^2 / λ^2

At 137kHz, the required $P_{ERP} = I^2 H_{eff}^2 /1687$
At 475kHz, the required $P_{EIRP} = I^2 H_{eff}^2 /100$

Example:
A 500W transmitter on 137kHz produces 2.4A antenna current in the Marconi antenna with H_{eff} of 11.2m in the previous example.

$P_{ERP} = 2.4^2$ x $11.2^2 / 1687 = 0.43$W.

We can also calculate R_{loss} and the antenna efficiency:
$R_{loss} = P_{RF} / I^2 = 500 / 5.76 = 87$ ohms,
Eff = 100% x $R_{rad}/(R_{rad} + R_{loss})$ = 100% x $0.041\Omega/(87 + 0.041)$ = 0.047%

121

Note that multiplying RF power by efficiency gives a radiated power of 0.24W, which is not the same as the power shown above. This is where the directional gain of 1.8 (over a dipole) comes in to make the ERP: 0.24 x 1.8 = 0.43W.

Calculating ERP for a loop antenna

$P_{ERP} = 1.8 \times 2 \times 320 \, \pi^4 \, I^2 \, A^2 / \lambda^4$
$P_{EIRP} = 2.52 \times 2 \times 320 \, \pi^4 \, I^2 \, A^2 / \lambda^4$

At 137kHz, this becomes $P_{ERP} = I^2 A^2 / (205 \times 10^6)$
At 475kHz, this becomes $P_{EIRP} = I^2 A^2 / (1.01 \times 10^6)$

Example:
The G3YMC loop in Chapter 3 has an antenna current of 8A at 137kHz and area of 100m².
Therefore $P_{ERP} = 8^2 \times 100^2 / (205 \times 10^6) = 3.1\text{mW}$.
The loss resistance of the loop is 0.65Ω, and the radiation resistance 27µΩ.
Therefore efficiency = 100% x 27µΩ / (27µΩ + 0.65Ω) = 0.0041%.

Why you need less RF power on 472kHz
The formulas for radiation resistance and ERP show that the power radiated for a given amount of antenna current is much greater at 472kHz than 136kHz, even taking into account the 'extra' 2.15dB produced by calculating EIRP on the higher band rather than ERP. This demonstrates that antenna efficiencies are much higher on 472kHz. Note also that loop antennas have a greater difference in efficiency than Marconis between the two bands.

How external factors affect radiated power
Experiments have shown that a low frequency antenna in an open field will outperform an identical one in a more enclosed space such as a suburban garden with shrubs, trees, fences and buildings, all of which will absorb low frequency RF. This means that any radiated power calculated from theory is likely to be higher than the real power by somewhere between 0 and 6dB. The actual amount is impossible to calculate, but at least it errs on the safe side so that you will be sure your radiated power will be legal if you do not exceed the calculated figure.

Field strength measurement

A more direct, and potentially more accurate, way of determining radiated power is by measuring the field strength at a known distance from the antenna. The relation between field strength E (volts/metre) at a distance d metres, and ERP is given by the formula:

$$P_{ERP} = \frac{E^2 d^2}{49}$$

For this relationship to be valid, the distance d must be in the far field region of the antenna, where the field strength falls away in inverse proportion to the distance from the antenna. The near field region is closer to the antenna, where the field strength decreases more rapidly with distance, and the formula above does not apply.

For small antennas at 136kHz and 472kHz, a safe minimum distance is about 1km. At distances much greater than a few tens of kilometres, the formula also becomes invalid, due to the effects of ground loss on the propagating wave close to the ground, and ionospheric reflections. For amateur stations, the signal is also likely to be too weak to accurately measure at larger distances.

Two pieces of equipment are required to measure field strength, a calibrated receiver and a calibrated 'measuring' antenna (pictured). The calibrated receiver must be capable of accurately measuring signal levels down to a few microvolts, and have sufficient selectivity to reject unwanted adjacent signals; the ideal amateur equipment for this purpose is the selective level meter (see chapter on receivers).

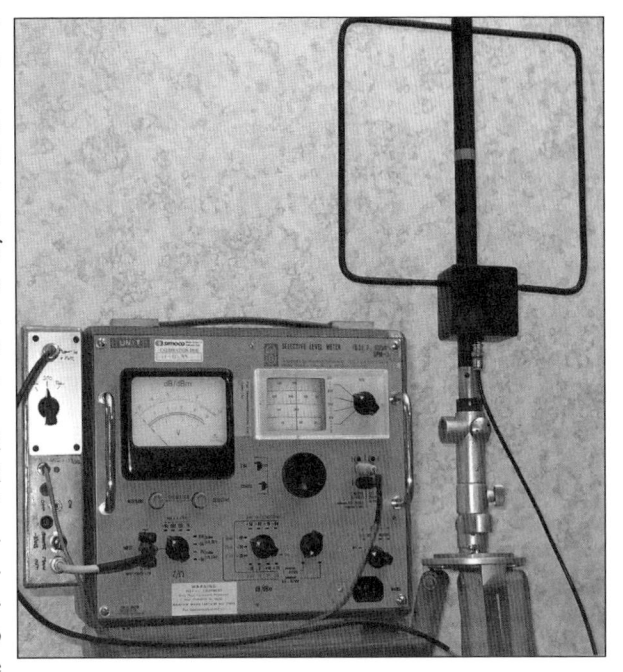

Field strength measuring system at M0BMU

Calibrated antennas have a specified antenna factor (AF), which is the number of decibels which must be added to the signal voltage measured at their terminals to obtain the field strength.

Quite good accuracy in the LF and MF ranges can be obtained using a simple single turn loop antenna. Such loops have a low feed point impedance, so the received signal level is little affected by the load impedance. The output voltage of an N-turn loop with area A square metres at a frequency f hertz is given by:

$$V = 2.1 \times 10^{-8} \times fNAE$$

From this, the antenna factor of a single turn loop is:

$$AF(dB) = 20 Log_{10}\left(\frac{1}{2.1 \times 10^{-8} \times f \times A}\right)$$

So at 137kHz, AF = $20Log_{10}(350/A)$
and at 475kHz, AF = $20Log_{10}(101/A)$

A square or circular loop made of tubing is usually used, with an area between $0.5m^2$ and $1m^2$.

As an example, suppose a signal level of 7.5dBµV (ie 7.5 decibels above 1µV, or 2.4µV; selective level meters usually give a decibel-scaled reading) is measured at a distance of 5km from the transmitting antenna, using a $1m^2$ loop at 137kHz. From the formula above, AF is 51dB, so the field strength is 58.5dBµV/m, or 840µV/m. Using the ERP formula gives P_{ERP} = 350mW.

123

A more compact alternative to the loop is a tuned ferrite rod antenna, however this requires calibration with a known field strength to determine the antenna factor. A field strength measuring system, including ferrite rod antenna, measuring receiver, and calibration set-up has been described by Dick Rollema, PA0SE [4].

Don't forget that the formula gives you ERP, which is OK for 137kHz. However, since the power limit is defined in EIRP (see above), it makes sense to use EIRP for your calculations on this band. To do so, simply multiply the resultant ERP figure by by 1.4.

Field strength measurements are prone to errors caused by environmental factors. The measured field strength is particularly affected by conducting objects giving rise to parasitic antenna effects. Parasitic antennas can be large steel-framed structures such as buildings and road bridges, overhead power and telephone wires, even such things as fence wires and shallow buried cables. These factors are difficult to avoid entirely, so several field strength measurements should be made at different locations over as wide an area as possible. Locations giving widely different values of radiated power can then be rejected; it will be found that a few decibels of variation still exists between different measurement sites, so the the final result should be taken as an average of several measuring sites [5].

The Scopematch tuning aid

This device developed by Jim Moritz, M0MBU, is used to simplify matching an LF antenna to 50 ohms. It is basically an SWR bridge without the detectors and meter. With this device the current and voltage amplitude and phase relationships can be monitored on a dual trace oscilloscope to establish a matched condition.

Construction

The circuit and construction is shown in **Fig 7.4**. Any high permeability 18mm diameter ferrite toroid core can be used for the transformers although a 3C85 core is ideal. The one in the prototype came from an old SMPS common mode choke. Do not use iron dust cores.

T1 secondary comprises 50 turns of enamelled wire. The primary is a single wire passing through the middle of the toroid, as used in an SWR bridge. Power handling is not really an issue.

T2 uses the same construction, but in this case the 50 turns are on the primary, with a single loop of wire to the output connection. Note that the 50 turn winding has the full transmitter output voltage across it, so the winding has to be well insulated from the core to withstand a few hundred volts of RF. The small core worked fine at the 400W level, but saturated with 600W. If high power is contemplated it would be advisable to use a larger core for T2, and to increase the primary and secondary turns from 1:50 to 2:100.

Operation

T1 is a 1:50 current transformer, which samples the current at the transmitter output and together with the 50 ohm resistor the scale factor is 1V = 1A. T2 is a 50:1 voltage transformer which samples the output voltage, 1V out = 50V at Tx output.

CHAPTER 7: MEASUREMENT AND CALCULATION

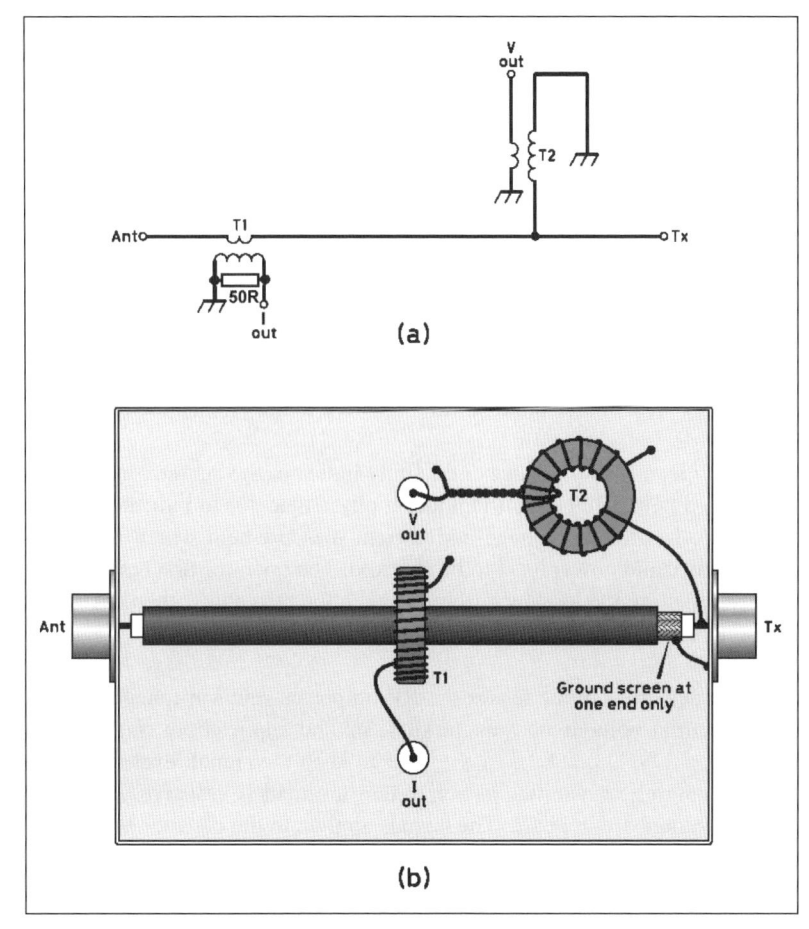

(left) Fig 7.4: The construction and circuit of the ScopeMatch. Note that the coaxial cable screen within the box is grounded on one side only

(below) Fig 7.5: (a) Perfect match. Current and voltage waveforms have the same phase and amplitude and only one trace discernible. (b) Antenna off tune and inductively reactive (voltage leads current). (c) Antenna resonant but resistive component of impedance low (voltage 25V, current 1A, 25 ohms) (d) Antenna resonant but resistive component of impedance high (voltage 50V, current 0.5A, 100 ohms)

The oscilloscope is set for the same volts per division on both channels. The scale factors are chosen so that when the antenna system is resonant at 50 ohms (or a 50 ohm dummy load is used), both voltage and current traces are identical (see **Fig 7.5**). If the load is inductive, the current waveform will lag the voltage; if capacitive it is the other way round. Getting the antenna resonant is just a matter of adjusting the loading coil until the two waveforms are in phase.

Once the antenna is resonant, if the current waveform is bigger than the voltage waveform, the load is less than 50 ohms, and if smaller the load is greater than 50 ohms. You can calculate the actual R by measuring V and I off the screen and using Ohm's law.

This gadget has proved very useful both for setting up an antenna and while operating it takes out most of the guesswork that occurs when using SWR bridge circuits. You can also tune up on low power.

125

T2 can be replaced with a capacitive divider. This comprises a 100p capacitor connected to the inner of the coaxial cable at the 'Transmitter' socket in series with a 5000p capacitor connected to ground. The centre of the capacitive divider is connected to the voltage-sampling socket. The 100p capacitor must be rated to take the transmitter output voltage.

Antenna tuning meter for 472kHz

The tuning meter system shown in **Fig 7.6** can be used with 472kHz transmitters with output powers of a few watts and above. A version of this circuit designed for higher power levels at 136kHz is described in [6]. When tuning up a transmitting antenna at LF or MF, the operator wants to know if the antenna impedance is inductive or capacitive, or tuned to resonance and, once resonance is achieved, whether the resulting resistive load matches the transmitter output.

The usual type of VSWR bridge circuit only indicates the degree of mismatch, so is of limited help. This circuit includes a phase detector to indicate resistive or reactive loads, and RF voltage and current measurement which allows the output resistance and power level to be checked. The operator first resonates the antenna by adjusting the loading coil inductance for zero phase, then checks the voltage and current levels to find the load resistance, and adjusts the matching circuit accordingly.

The phase detector circuit senses the RF output current via transformer T1, and the RF output voltage via inductors L1, L2 and applies both signals to the diode ring D5 - D8. D3, D4 act as a clipper to keep the signal level fairly constant with changing transmitter power. When the load is resistive, the output current and voltage are in phase. The signals applied to the detector are approximately in quadrature due to the inductive voltage divider. This results in a DC output from the phase detector close to zero. A reactive load results in the RF voltage and current no longer being in phase, and a positive or negative deflection of the centre zero phase meter depending on whether the impedance is inductive or capacitive. Due to phase offsets and diode mismatches within the circuit, the phase detector output is not quite zero with a resistive load; R4, R7

Fig 7.6: 472kHz antenna tuning meter. Component details are shown in Table 7.1

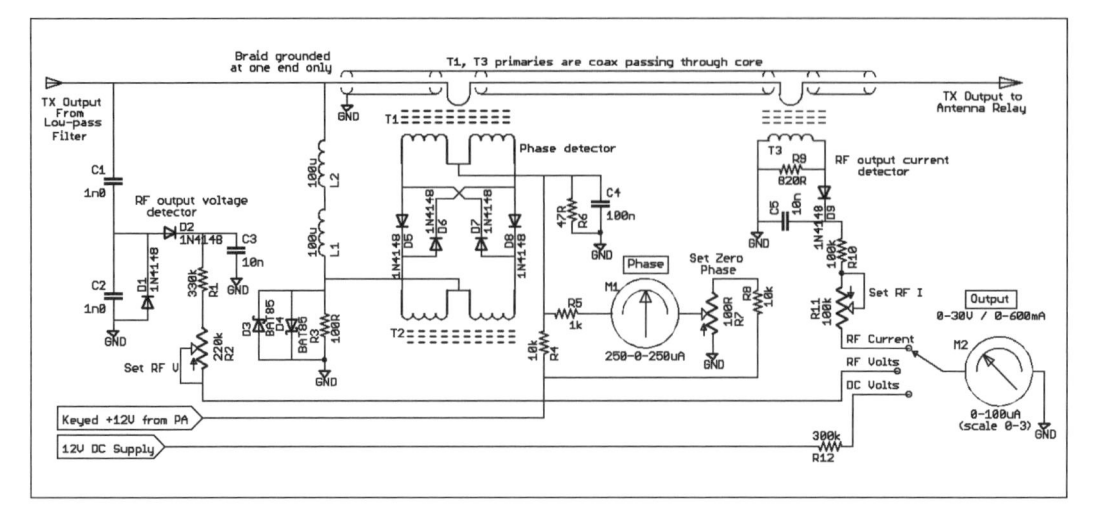

CHAPTER 7: MEASUREMENT AND CALCULATION

and R8 provide an adjustable DC bias to allow the meter deflection to be set to zero with a dummy load attached to the transmitter output.

The RF voltage is sensed via a capacitive divider C1, C2. The reactance of C1, C2 cancel the reactance of L1, L2, minimising the loading effect of the meter on the transmitter. A diode voltmeter produces a DC voltage to drive the meter. The RF current is sensed via current transformer T3, and applied to a second diode voltmeter circuit. The scale-setting resistors are chosen so that the voltage scale is 50 times the current scale; thus, when the load is 50 ohms, the meter deflection is the same on both voltage and current ranges, and the operator can see at a glance the load is matched, greater than, or less than 50 ohms. A third meter range is provided to measure the DC input voltage, for use with a battery supply.

> C1, C2 are 1nF polystyrene, polypropylene or silver-mica.
> L1, L2 are Axial chokes, 100uH.
> T1: secondary is 2 x 20 turns bifilar 0.25mm wire on 14mm 3C85 ferrite toroid, and the primary is RG58 coax passing through the core.
> T2 is 2 x 20 turns bifilar 0.25mm wire on 14mm 3C85 ferrite toroid.
> T3 secondary is 50 turns 0.25mm wire on 14mm 3C85 ferrite toroid, and the primary is RG58 coax passing through the core.

Table 7.1: Tuning meter component notes

The scale-setting resistors are chosen to suit the available meter movements and scales. R5 sets the phase meter sensitivity to provide near full-scale deflection with the maximum power level, but an accurate calibration is not required since the phase meter only has to indicate zero, positive and negative. Operation of the phase meter can be checked by connecting a capacitor of 5 - 10nF across the dummy load, which should produce near full-scale deflection. R1, R2 set the voltmeter scale, R10, R11 set the RF ammeter scale. The meter can be calibrated using an oscilloscope or known RF voltmeter or ammeter with a dummy load. Accuracy is not critical; the main thing is that the meter registers equal deflection on voltage and current ranges with a 50 ohm load.

Frequency stability and calibration

In order to receive stations using QRSS at the slower speeds, such as QRSS30, QRSS60, QRSS120, and some data modes used for DX working, it is necessary to have a very stable radio. Fortunately, most modern receivers are good enough. However, a means must be found to ensure that the waterfall display on the receive software screen is looking at the precise frequency that you want. Remember that the waterfall is monitoring an audio frequency from your radio. Fortunately, there are fairly simple ways of doing this without expensive test gear. The description below assumes you are calibrating for QRSS operation using *Argo* (see the Modes chapter), but the basic idea will work on other software employing waterfall displays.

Current versions of *Argo* and *Spectrum Lab* have built-in utilities to facilitate calibration of sound-card frequency errors. The resulting calibration factors are then used to automatically correct the spectrogram frequency scales. This should be done before the steps shown below.

The LF spectrum is used for some highly accurate transmissions such as the time-standard GBR at Anthorn on 60kHz and BBC Radio 4 at Droitwich on 198kHz. Other standards are available in the USA and elsewhere.

The calibration method is as follows. First set your LF receiver to USB and tune to one of the accurate stations, with the radio's frequency readout displaying the exact frequency, eg 198.000kHz. Set *Argo* to CW and find the station on the Argo screen - it will appear as a thick horizontal line. Note the audio frequency displayed by Argo. Reduce the receiver's RF gain so that the Argo displays a thin line. Set Argo to QRSS3 and adjust its audio centre frequency to that noted in the previous step. Reduce the receiver gain and alter the Argo frequency until the station displays a thin horizontal line roughly half way up the screen. Next, set *Argo* to QRSS10 and repeat these steps, gradually reducing the *Argo* speed until you reach the desired dot length. This can take a while as the display slows down considerably at the longer dot lengths.

Once the commercial station is drawing a thin line in the centre of the QRSS120 (or whatever) screen, use the calibration facility in *Argo* (from the set up menu) to set the audio frequency (displayed next to the line) to zero. Calibration can be fiddly, but is worthwhile. Now you know that the frequency displayed on your receiver equates to the zero on the screen. A signal that appears 0.1Hz above the zero line will be 0.1Hz higher than the frequency displayed on your radio; the combination of the receiver and *Argo* has given you a much more accurate frequency display than the receiver alone.

Now set your receiver so that it displays the frequency you want to monitor on the amateur band, turn up the gain and you are ready to receive. Note that instead of zero, Argo can display a frequency that ties in with the wanted frequency in the band, eg display 989.0 for 474.989kHz. The calibration can be checked at any time by switching your receiver to 198.000kHz and turning the gain down.

Note that calibrating the receiver at a frequency that is significantly different from the amateur band frequency of interest may result in a significant inaccuracy if the receiver's oscillator is subject to relatively large errors. This can be minimised by using an off-air calibration signal closer to the desired receive frequency, for instance the DCF39 "carrier" on 138.830kHz, which has adequate precision to calibrate the receiver to within about 0.1Hz.

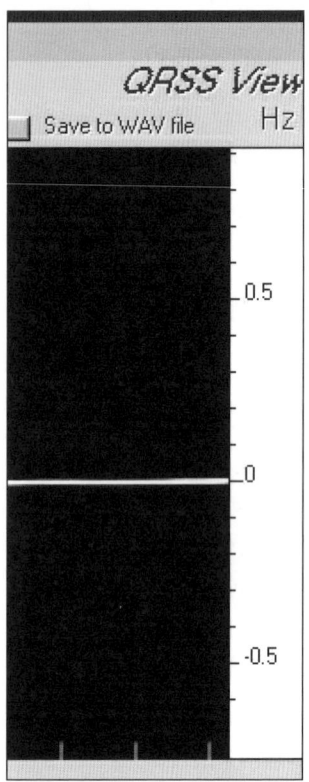

Detail of *Argo* screen showing the result of calibrating for frequency resolution of better than 0.1Hz

References

[1] *http://www.arraysolutions.com/Products/AIM4170.htm*

[2] *http://uk.rs-online.com/web/* - RS Components on-line catalogue

[3] *http://www.ifwtech.co.uk/g3sek/clamp-on/clamp-on.htm* (note that a suitable core must be used for LF use)

[4] Field strength meter: *http://www.wireless.org.uk/pa0se.htm*

[5] 'Experimental investigation of very small low frequency transmitting antennas', J.R. Moritz, IEE 9th International Conference on HF Radio Systems and Techniques, June 2003, *IEE Conference Publication n. 493* pp 51 - 56.

[6] LF tuning meter: *http://www.picks.plus.com/software/LFtunemeter.pdf*

8

Low frequency propagation

by Alan Melia, G3NYK

In this chapter:

- Ground waves
- Sky waves
- The effect of the Sun
- Can we predict good conditions?
- Fading
- Longer term changes
- 472kHz

IT IS GENERALLY BELIEVED that propagation in the low frequency bands is stable, almost boring. This is only loosely the case. Historically low frequencies have been used where a wide service area is required which is relatively immune from interference and ionospheric effects such as fading. Prime commercial uses have been radio navigation systems, wide area military communications, and to a lesser extent, mainly in Europe, broadcasting. A lot of the research that has been carried out has had as its objective a determination of the severity of interference from signals beyond normal service range. In the years after WWII this was more to determine the reliable range of long distance navigation systems such as Decca Navigator, Loran-C and Omega. The critical factor being the reliability with which the phase or timing of the signal could be determined.

By contrast, the radio amateur, often working at the threshold of possibility, is willing to wait for and use whatever short-term effects are available to achieve his ambition of long distance communication.

The signal from a transmitter may reach a receiving site in two ways. Firstly by way of waves which follow the curvature of the Earth to some extent, known as ground waves. Secondly by the return of skyward travelling waves by the ionosphere, referred to colloquially as skywaves, or more correctly ionospheric waves.

Ground waves

The so-called ground waves follow the curvature of the Earth because the speed of the wave is slowed slightly by the dielectric constant of the ground (**Fig 8.1**). This has the effect of tilting the wave-front downwards, and allows the signals to be detected far beyond the normal visible horizon.

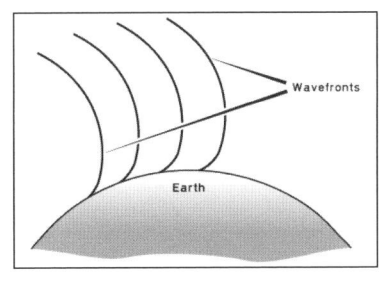

Fig 8.1: Ground wave signals are tilted downwards, allowing them to 'hug' the earth and propagate far beyond the visible horizon

LF TODAY

Unlike higher frequencies the strength of the ground wave signal is not reduced significantly by absorption. As a result there is no 'dead zone' on low frequencies and ground wave signals can be detected at over 2000km from the transmitter.

Sky waves

Because most amateur sized aerials are small compared to the wavelength, a considerable amount of the radiated power is launched at higher angles and rapidly leaves any influence of the ground. These waves travel upwards until they reach the ionosphere at around 50 to 100km altitude (**Fig 8.2**). Vertical incidence signals will penetrate deeply into the ionised regions but will suffer a great deal of attenuation. At lower angles the waves will be gently 'bent' (refracted is a more accurate term) back towards the ground. Skywave returns have been detected at as little as 300km from the transmitting station and result in a slow shallow fading in the strength of the signal.

The change in strength is due to the change in the distance the skywave travels as the altitude of the 'bending' region alters. The skywave arrives at the receiver with a different phase to that of the ground wave and the two waves may either add to reinforce the signal, or cancel to reduce it.

Complete cancellation only occurs if the ground and skywave are the same strength as well as 180° out of phase. Most of the published data suggests that the skywaves become approximately equal in strength to the groundwaves at around 700km from the transmitter. Beyond this distance the skywave is stronger.

A case of 'dead zone' can appear when very low power signals are transmitted. In this situation the ground wave is weakened, by the nature of its outwards spread, to levels below the detection level of the receiver before the angle of the skywave becomes low enough to cause them to return. This is often experienced by US FCC Part 15 stations (operating between 160 and 190kHz), who are limited to one watt RF input power and a maximum antenna length of 50 ft.

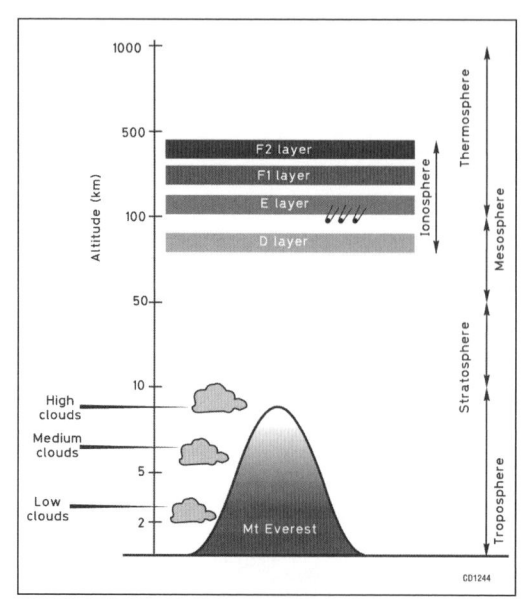

Fig 8.2: The ionospheric layers. Note that their height will vary with the time of day. LF signals are propagated by the lowest part, the D layer

A simple geometric construction (**Fig 8.3**) allows us to calculate the distance covered by a single ionospheric 'bounce' provided we know the height at which the signal is bent back towards Earth. For simplicity we can consider a mirror like reflection from an altitude we will call the 'apparent reflection height' and we will assume that the signal leaves the transmitting site tangentially to the ground.

Experience suggests that the daytime 'reflection level' is around the bottom of the D-layer at about 50km alti-

CHAPTER 8: LOW FREQUENCY PROPAGATION

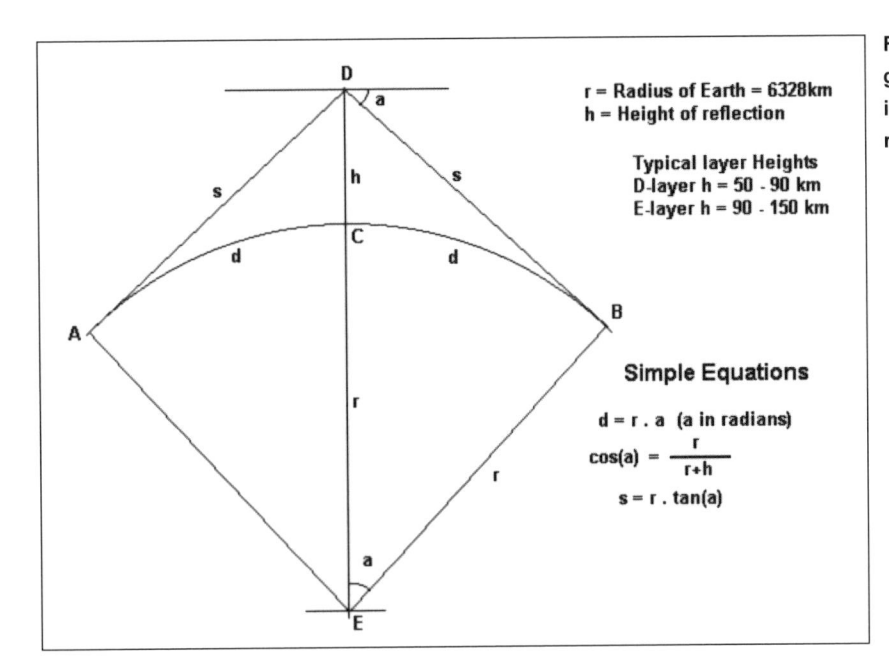

Fig 8.3: Simple geometry of ionospheric reflection

r = Radius of Earth = 6328km
h = Height of reflection

Typical layer Heights
D-layer h = 50 - 90 km
E-layer h = 90 - 150 km

Simple Equations

$d = r \cdot a$ (a in radians)

$\cos(a) = \dfrac{r}{r+h}$

$s = r \cdot \tan(a)$

tude whilst at night the reflection level is in the upper D-layer near the bottom of the E-layer at around 100km altitude.

Our calculations then show that in daytime a single hop will be of around 1000km, whilst at night-time a single hop will be around 2000km. It is important to realise that the returning skywave approaches the ground at grazing incidence or tangentially, it does not bounce at high angles like a tennis ball as shown on many sketches. Thus the wave does not need to be 'reflected' from the ground to go upwards for a second hop, it merely slides past barely touching the ground.

Low frequency radio paths can comprise several such 'hops'. The regular 136kHz signal heard from VO1NA during 2003 and 2004 at 3600km was probably a two hop path, whilst the record contact between Quartz Hill in New Zealand and Vladivostok in Asiatic Russia at over 10,000km was probably around five hops. One-way signals from Quartz Hill were detected and identified in western Russia at 16,000km. These exceptional distances were achieved at night with the path in full darkness.

The daytime path lengths are usually restricted to around 2000km, due mainly to the higher signal absorption (attenuation) levels in daytime and the loss at each 'hop'. Nevertheless, under exceptional circumstances VO1NA has been copied in the UK at 1200UTC, but this is a rare event. Even the powerful (20kW ERP) naval station CFH at Halifax, Nova Scotia, is not often heard in Europe during the day-time.

In practice the situation is a little more complicated. The 'reflection' of the waves is not without loss. At lower altitudes the daytime ionisation of the D-layer produces a belt of ionisation below the 'reflection height' that absorbs power from the radio waves. At the lower 'HF' frequencies (160m, 80m and 40m) this shows up as the severe restriction in day-time range, because the skywaves are completely absorbed by the D-layer. After sunset this ionisation in the

D-layer decays and the lower HF waves can pass through to be reflected from the F layer and so called 'skip' signals appear. In the LF frequency range the daytime absorption is not so complete and it is possible to receive daytime skywave signals at distances of about 2000km, which probably requires two hops. For a given path these signals are never as strong as those received after dark.

The effect of the Sun

Propagation conditions are further affected by solar disturbances. At the height of the sun spot cycle, the Sun emits bursts of intense X-rays and ultra-violet light called solar flares. Because the output of a solar flare is an electromagnetic emission (like a radio wave), it is almost instant and will only affect the side of the Earth that is currently illuminated. Intense flares cause radio blackouts at HF frequencies, because they produce extra ionisation in the D-layer which strongly absorbs most HF radio frequencies (**Fig 8.4**). Surprisingly at LF the effect is usually the exact opposite. The intense radiation converts the normally absorbing level ionisation to a state where it easily 'reflects' LF waves. The result is that LF signals show a strong peak in strength which is a similar shape in time to that of the X-ray flux (as can be seen on the NOAA web site [1]).

This flare enhancement can cause an increase of up to 10dB in the strength of a signal being received in daytime. This can be useful, but the enhancement is much less than is normally achieved on night-time paths.

The solar magnetic disturbances that produce flares also throw off huge clouds of ionised gas or plasma. The plasma travels much more slowly and takes between 36 and 56 hours to reach the vicinity of Earth, a journey of 96 million miles. When these clouds reach the earth they buffet the atmosphere. Because they are composed of fast moving charged particles, the cloud carries with it a magnetic field, and its interaction is with the Earth's magnetic field. The magnetosphere is a distorted doughnut shaped 'cage' formed by the lines of magnetic force generated in the Earth's core. The magnetosphere protects us from most

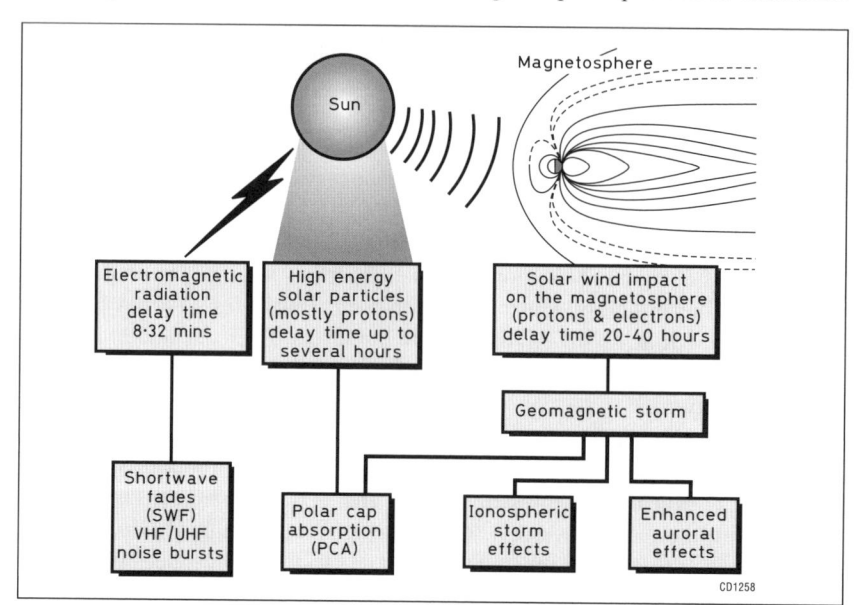

Fig 8.4: Effects of solar disturbances on the ionosphere

of what the Sun can throw at us. Without it, the majority of life on Earth could not exist.

If the magnetic field of the plasma is in one direction the cloud bounces off fairly harmlessly, like similar poles of small bar magnets. In the opposite field direction the field lines of the plasma are said to 'connect' with the geomagnetic field, like the bar magnet opposite poles. This situation opens up 'cracks' in the Earth's 'defences' and charged particles flood into the atmosphere. The most notable visible effects of this phenomenon are the aurora seen mainly in high latitudes after a magnetic storm.

The event also causes the Earth's field to vary wildly for a short period, an event which can be observed on magnetometers, and hence the term 'Geomagnetic Storm'. This was though to be the main source of injected particles, but since then the use of satellites and the discovery of the Van Allen radiation belts has refined ideas of the process. It has been realised recently that the majority of the particles are swept past Earth and are sucked into the long tail of the magnetosphere on the side of Earth opposite to the Sun. The Geomagnetic field then draws them back into the Van Allen belts, forming a series of circulating rings.

Electrons travel one way and ions, because of their opposite charge, the other. More than this, the electrons being much lighter follow paths that spiral round the Earth's magnetic lines of force from one hemisphere to the other. They are 'turned round' by the 'cramping' of the lines of force near the poles. Together these rings of circulating charges, which are known as the Equatorial Ring Current, generate a magnetic field of their own, which can be detected and measured by magnetometers at the Equator. The Rings exchange charged particles, mainly electrons, with the ionosphere.

It had been noted by many researchers prior to the advent of satellite measurement, that the injection of electrons (called electron precipitation) into the ionosphere, after a geomagnetic storm, led to severe reductions in distant radio-signal strengths. The effect did not build up until a day or so after the storm and could often persist for up to 28 days. It is not physically possible for electrons, even very energetic ones, to exist in the relatively high atmospheric pressure at the D-layer for very long, so the signal attenuation should be expected to decay with the passing of the storm. It has recently become clear that the Ring Current acts as a reservoir of electrons which are bled into the ionosphere at the daylight edge, where the magnetosphere is distorted by the pressure of the solar wind. Thus the after-effects of a Geomagnetic storm on LF radio transmissions will be felt until the Ring Current is depleted of its trapped electrons.

Fortunately the Ring Current can be measured by the field it generates. It is not an easy task because the field due to the Ring Current is about one thousandth of the Earth's field (50 to 400nT against 50,000nT).

Daily estimates of an index,which effectively measures the Ring Current, are published by several institutes. Most useful for LF radio propagation prediction are the (hourly) real-time estimates from Colorado University and Kyoto University, which are both available on the Internet [1, 2]. The Index is referred to as 'Disturbance Storm Time' and carries the mnemonic Dst.

Plasma clouds (referred to on solar web sites as coronal mass ejections or CMEs) can also be produced by disturbances in the solar atmosphere known as

LF TODAY

A very useful indication of ionospheric propagation at LF can be derived from the data provided by Colorado University [2]

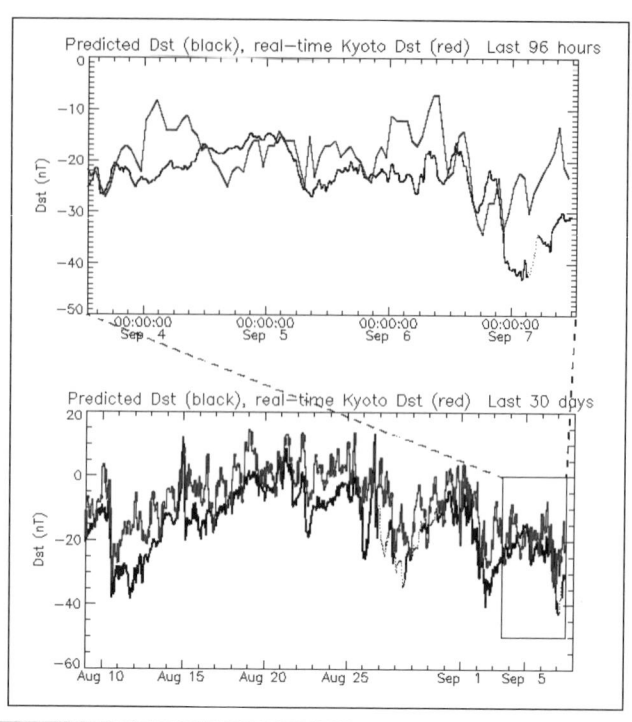

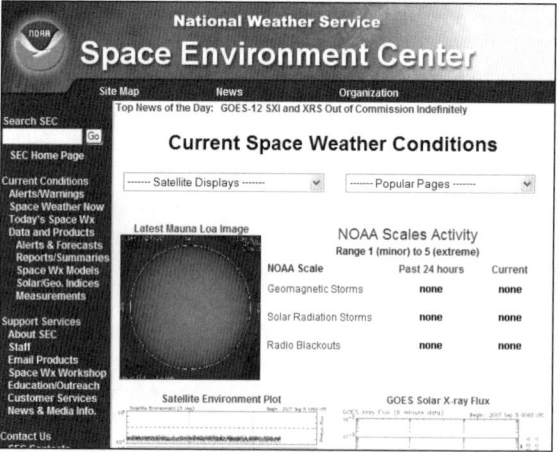

(left) Check the NOAA SEC web site [3] for solar flare information

(below) Dst readings are available on the web site at Kyoto [3]

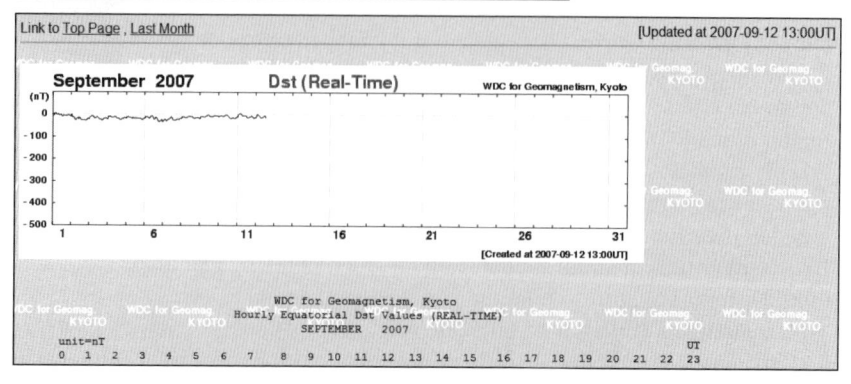

134

coronal holes. Whilst flare-associated events are more prevalent in years of high sunspot activity, coronal hole events occur throughout the solar cycle including periods when the visible face of the Sun is totally devoid of spots.

The most familiar effect of these disturbances is intense aurora. A rarer, but more serious problem, is that these events can induce massive currents in long northern power distribution systems. Canada suffered a power black-out for several hours some years ago due to one such event. Submarine cables and satellite communications systems can also be disabled. The arrival of a CME can herald the onset of a period of poor HF communications. Absorption, due to the enhanced ionisation of the D-layer by the trapped electrons, causes all the bands to go 'flat'. Again the effect at LF is different. The daytime signals up to about 2000km can be significantly enhanced by up to 10 or 12 dB above normal levels. The effect on night-time paths is more dramatic with absorption at LF increasing significantly above normal and signal levels on long paths dropping as much as 20dB below normal levels.

Can we predict good conditions?

It might seem that with the knowledge built up it should be possible to predict good LF conditions. This can be done to some extent, but it is much easier to predict bad conditions particularly for night-time paths. Flares cannot easily be predicted and the best that NOAA will say is that there is a likelihood of a flare of particular strength in the next three days. However, by carefully watching the background levels via the NOAA SEC internet site [1], it is possible to see the Solar X-ray flux, measured by the GEOS satellites, increasing. This gives a good indication that a predicted flare is imminent. Flares are more likely around the sunspot maximum and their likelihood is directly related to the sunspot number, which is a measure of solar activity.

Coronal mass ejections are detected as they form and their impact on Earth can be predicted about two days ahead of their arrival, which is denoted by a large increase in the geomagnetic index Kp, known as a geomagnetic 'storm'. One problem is that a CME will sweep past Earth, the Kp will indicate the storm and return to the 'quiet' state in a matter of hours.

The effect of the storm is not felt in radio terms for about two days at mid latitudes, despite aurora occurring on the night of the impact. This may be due to the time taken for electrons trapped in the ionosphere to diffuse down and spread out at lower latitudes. It may also be due to the fact that the charged particles are collected in the magnetosphere 'tail' as the plasma cloud sweeps past the Earth, and these must travel back, under the influence of the magnetic field, into the Ring Current. Then 'precipitation' from the Ring Current takes place mainly at the 'dawn edge'.

An intense storm will produce a depression in night-time signal levels that can last for 21 to 28 days and the Kp index is no indicator of the return of good conditions but a study of Dst during this period is more rewarding.

A correlation between this index and signal levels of well known commercial stations suggests that the Dst index mirrors the return of good radio propagation. The Dst index has a range from small positive values (0 to +40) for quiet conditions to -100 to -400 for intense storm conditions. The units are nanotesla (nT) because this is a magnetic field. Unlike the Kp index it will show the effect of small events following a major storm which tend to extend the recovery time to good radio conditions, because of the 'reservoir' being topped up.

The relationship between the Dst index and the precipitation of electrons into the ionosphere is described by Hargreaves [4] and Daglis [5].

The extra attenuation of LF signals after a geomagnetic storm is reported by Jack Belrose, VE2CV, and others in learned papers in the 1960s. Although the attenuation can be observed on shorter night-time paths at less than 1500km, it is more obvious on the longer multi-hop paths.

The plot in **Fig. 8.5** is a re-analysis of data taken in 2003 of the strength of the Canadian Naval station CFH on 137.0kHz in the UK. This path is 4500km and so is predominantly 2 hops. The Path Index has been defined as the number of seconds that the signal in the UK exceeds a given level. For the benefit of this plot the level is 32dBµV or 40µV. This is a high level, but CFH has an ERP of around 20kW. The cumulative period that the signal exceeded 32dBuV has then been adjusted to allow for the period that the path was in darkness. Thus the ordinate is a fraction of the available time the path was open.

It can be seen that there are no usable periods when the Dst in the preceding day was lower than -60nT. If the Dst was above -60nT the likelihood of a usable path increases as the Dst rises to -20nT. This is the Dst level during a normal quiet period when the equatorial ring current is depleted to levels that are supported by the normal solar wind flow.

Fig 8.5: Plotting periods of 'good' transatlantic propagation versus Dst shows that it is easier to predict poor conditions than good

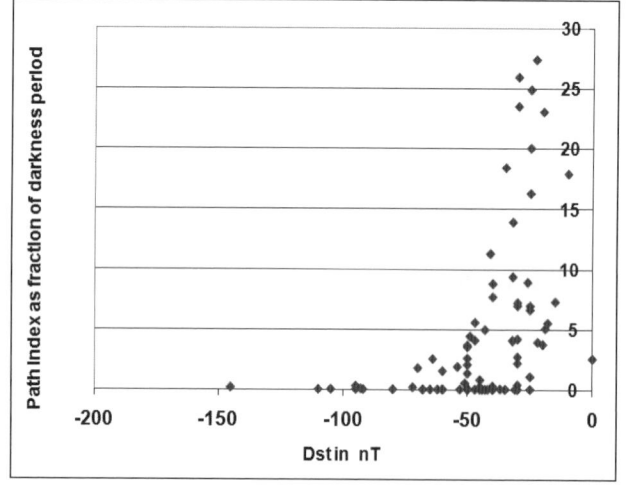

The plot supports the subjective experience that good DX is most likely to be worked when the Dst is around -20nT or higher. However it raises another question. There is not a linear relationship between the Dst and the Path Index so there must be some other mechanism in play.

Fading

Fading is a phenomena well known by all radio users. It is manifest by the signal being received rising and falling in amplitude. This implies that the signal level at the antenna terminal of the receiver is rising and falling. If the signal were reaching the receiver by a single path there are no mechanisms that could cause these drastic strength changes. Thus the cause of fading is the reception of the signal over two slightly different changing paths.

At low frequencies the groundwave is propagated well and can still be as strong as the skywave at a range of 700km. At this point the received signal is the summation of the two waves in the receiver. However the strength of the

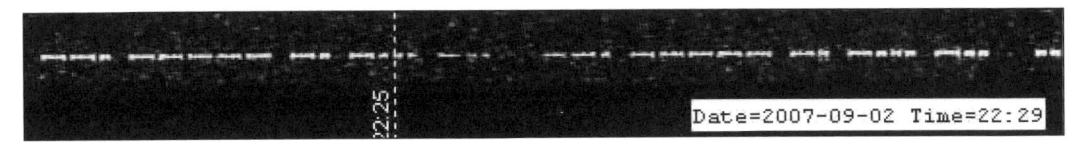

CHAPTER 8: LOW FREQUENCY PROPAGATION

Fading on MF is illustrated by this ten minute section of G0NBD's 500kHz QRSS3 transmission received over a path of a few hundred kilometres by Hartmut Wolff in Germany

signal received is dependent on the relative phases of the two waves. The ground wave will produce a steady signal of constant phase, but the phase of the skywave will depend on the total path length through the ionosphere. If the ionosphere is stable and the apparent reflection height does not change, a signal of steady strength will be received. If however the height of the 'reflection' changes, as it usually does during the darkness period, the phase of the skywave will change relative to the ground wave. This produces the familiar slow cycling of the received signal strength. If the two signals are in phase an enhancement of up to 6dB is achieved, but if the two signals are in anti-phase a deep dip is heard with the resultant signal being the difference in amplitude of the two.

Fading occurs on long distance paths too where the groundwave is so weak as to be negligible. Then the fading is the result of the summation of different skywave paths, for example a two-hop path and a three-hop path. The three-hop signal will usually be weaker than the two-hop because it suffers attenuation at each pass through the ionosphere. It is probable that this is what accounts for the 'poorer-than-expected' performance on nights when the Dst is around -20nT.

On the CFH to UK path, the predominant mode is two-hop because the path is 4500km and the maximum hop length at LF s around 2000km. However one-hop signals are present, shown by the increase in strength when the shadow reaches mid-Atlantic at 100km altitude. It is most probable that three-hop signals are also present when the attenuation is low. We now have three signals combining in the receiver but we do not have one with a constant phase, they are all changing as the ionisation in the night-time ionosphere changes. It is possible to envisage that the one-hop and three-hop signals could interact with the main two-hop signal that would give long periods of poor reception. Unfortunately this cannot be checked with CFH because it would require the phase of the received signal to be measured, and the modulation on that transmitter makes it impossible. Overnight phase measurements have been made by amateurs on 136kHz on a 1200km path over 48 hours and phase changes of around 1200 degrees were observed overnight.

Longer term changes

Radio amateurs have had access to LF for almost a whole solar cycle now, and we can say what the effect of the 11 year solar cycle on LF conditions might be. The cycle is usually defined in terms of sunspot numbers and solar flares will be directly related to the sunspot number. Geomagnetic events seem to peak a year or two past the sunspot maximum. Geomagnetic storms still occur during the quiet years but the period between them becomes longer allowing more time for propagation conditions to recover. Perversely it would seem that the very best results are not achieved in dead quiet solar conditions and a small amount of geomagnetic activity can be an advantage.

We can now see that the higher level of daylight solar flux during the years around the maximum produce stronger daylight skywave signals. On average the daylight path of in excess of 1000km, can be 10dB lower at the Solar minimum.

There is a further effect which cannot yet be quantified which shows that night-time signals seem to decline below expectation levels in a long geomagnetically quiet period in the solar minimum. This may be due to the lack of sufficient radiation to retain the necessary level of ionisation in the lower E-layer. In effect the E-layer may be becoming slightly transparent to 136kHz signals, and some is 'leaking away'. It is noticeable that there can be a dramatic recovery to good levels following immediately after a small geomagnetic disturbance which only measures at a Kp of 4. During a long quiet spell in January 2003 a minor geomagnetic storm raising Kp to only 4, produced record levels on transatlantic paths.

There is a mythology that LF propagation only exists in the winter months, between the autumn and spring equinoxes. This is not true, daytime levels are better in summer because the sun is stronger at mid-day and the ionisation for daytime skywave is higher. The darkness period is shorter and some paths may not have a period of total darkness for a month around the summer solstice. What normally limits operation in summer is the level of static, interference from lightning crashes. This makes listening very uncomfortable and tiring, but some data modes and QRSS can be operated successfully through this period on nights when the interference is not continuous.

The best information about the state of propagation is derived from listening to reliable commercial stations and logging their strength. Propagation is very frequency dependent and it is only really useful to listen to stations in the same frequency area. The frequency band from 75 to 200kHz is totally different to the bands above and below, so suitable indicators for 136kHz conditions should be chosen from that range.

This runs from the Frequency Standard transmission HBG on 75 kHz in Switzerland, to BBC Radio 4 on 198kHz. Choose a station that is at least 1500km away so that you are receiving only skywave at night. BBC Radio 4 and the German utility station DCF39 on 138.83kHz are too close to be useful as propagation indicators in the UK.

CFH in Nova Scotia on 137.00kHz (as used in the analysis above) is ideal at a distance of about 4500km from the UK, but it does have long periods of inactivity. Greek naval station SXV located near Athens on 135.75kHz is one choice at a path length of 2200km from the UK, but it is difficult to understand the strength variations.

This may be due to the long reach up the Adriatic followed by a passage over the Alps, but it is also suspected that the power is reduced at times. Under generally good conditions the received level from SXV can swing violently up and down, probably due to multi-hop fading.

Because fading is a function of the path distance, it can be very dependent upon the location of the receiving station. Two stations just 50 miles apart may experience what seems like totally different conditions, if one is ideally situated to benefit from constructive (additive) interference between paths with numbers of different hops.

472kHz

Much of the above LF propagation information is also applicable to the allocation at 472kHz. The major differences are, of course, a function of the higher frequency. Radio amateurs have had a full solar cycle of experience on 136kHz but MF effects are relatively new to us.

Radio propagation at 472kHz is subject to the same effects as signals at 136kHz, though there are quite different reception characteristics due to the shorter wavelength. The interaction of the 472kHz signal with the ground conductance means that the groundwave does not travel as well and is relatively weaker than a 136kHz groundwave at the same range. At short distances the 472kHz signal is often subject to fast and deep fading because the skywave is a more important part of received signal. This can mean that chunks of say a slow QRSS signal can be lost at slower dot-rates. The speed of the fading is due to the fact that the wavelength is much shorter and small changes in the ionosphere of the 'apparent reflection height' can result in large changes in phase with respect to the ground wave.

Fading over longer ranges is less troublesome because the groundwave diminishes and the fading is then due to interaction between different propagation modes, for instance one-hop and two-hop signals. These paths are both changing in length with any movement in the 'reflection height'. This can mean that poor propagation conditions persist for extended periods, but conversely good conditions can persist too. This effect can be very dependent on path length, so like the HF bands if someone else is having a difficult time the path might just be right for you.

As with 136kHz the ionospheric return is from the lower E-layer (90 to 100km altitude) at night so one-hop range is similar at around 2000km. The more hops a signal requires to take to cover the range the more attenuation it will suffer on nights when the Dst is below its best.

When timing long paths for the length of darkness, do remember that the path ends do not require to be in ground level darkness. Provided there is darkness at 100km altitude 1000km down range of the transmitter, the first hop is possible. Also the last hop will probably reach the receiver anything up to an hour after dawn or before sunset. This gives a potential 2000km extra range over the pure darkness path.

Continuous propagation monitoring is more difficult around 472kHz as surprisingly there are few usable or suitable propagation beacons, close to the allocation. The bottom end BC stations share frequencies and change power from day to night. The Non-Directional Beacons (NDBs) are relatively low power (100W) so often cannot be heard at extreme range over the full 24 hours. They also share frequencies so it is sometime difficult to detect which site you are receiving on an unattended data logger. The use of exact carrier frequencies can help here, provided that there are no local stations on or near their frequency.

References

[1] Colorado University Dst data: *http://lasp.colorado.edu/space_weather/dsttemerin/dsttemerin.html*

[2] Kyoto University Dst index page. *http://wdc.kugi.kyoto-u.ac.jp/dst_real-time/presentmonth/index.html*

[3] National Oceanic and Atmospheric Administration (NOAA). *http://sec.noaa.gov/*
[4] *The Solar Terrestrial Environment*, J R Hargreaves, Cambridge University Press, 1992
[5] 'The Terrestrial Ring Current: Origin,Formation,and Decay', Daglis, et al, *Rev. Geophys*, 37(4) 407-436, 1999

9

Operating practice

In this chapter:

- Operating on 472kHz
- Operating on 136kHz
- QRM and QRN
- Remote receiving/grabbing/reporting
- DX Working
- Operating away from home

THE MAIN THING that differentiates the low frequency bands from most other amateur allocations is their tiny size. The 136kHz band is just 2.1kHz wide and the 472kHz band is not much wider at 7kHz. Although various modes are permitted in the UK, it is obvious that anyone using conventional speech modes would be unpopular.

As with any band, the key to success is to listen first. Listening will give you an idea of what stations are available and when, what frequencies and modes are common and what conditions are like from day to day, and from daytime to nighttime. Many operators appreciate reception reports, particularly from those new to low frequencies, and nowadays there are some easy ways to deliver reports in real time via the internet, so there is no need to be in too much of a hurry to start transmitting.

If you have access to e-mail, you can join the RSGB LF Group [1] and become part of the experimental community; see Appendix 1 for how to do this.

Because the experimental nature of the low frequency allocations, activity levels are lower than on HF. However, there is activity most evenings and every weekend, especially during the winter months. If no signals are heard, do not assume that there is no-one on the band. Often there are people monitoring for activity who will respond to a CQ call, even during the working day.

Most contacts take place co-channel, that is both stations are on the same frequency. However, occasionally split frequency operation is used. It may be because one or both stations are crystal controlled, or to avoid DX reception being swamped by locals. If you don't get a co-channel reply to a CQ call, it may be useful to tune the band for any other callers; and on these narrow bands this doesn't take long! Cross-mode skeds, for example between CW and QRSS, should be set up with each station in the appropriate part of the band.

A CW contact exchange is very much the same as on the HF bands. This ranges from the 'rubber stamp' contact - often when signals are weak or

LF TODAY

Gary Taylor, G4WGT in his shack. The equipment includes:
(top shelf) 150W MF amplifier module
(2nd shelf) G0MRF 300W LF Tx, DDS and scopematch
(3rd shelf) Audio test gear
(4th shelf) LF and MF SDRs, LF/MF linear transverter, NRD-345 receiver
(bottom shelf) External soundcard
(desk top) TS-440 transceiver used as transverter drive, laptop PC
(not shown) high performance PC running multiple grabbers

between stations who do not share a common language - to the 'ragchew' which is just a normal conversation between two friends.

An exchange of information differs from that given at HF in the addition of a 'locator'. Like the bands above 30MHz, distance is important on low frequencies, so the IARU Locator (usually called 'Loc') is often sent during the first contact between stations or in some beacon transmissions.

If you are unfamiliar with this grid system, or need to work out your own locator, full details can be found in the RSGB Yearbook [2]. Alternatively, a web site containing a calculator that will convert latitude and longitude into a locator and even calculate the distance between you and the station you worked, is at [3]. There are also apps for Android phones/tablets and the iPhone.

Operating on 472kHz

During daylight, propagation is mainly via ground wave and results in reliable signal levels at distances up to a few hundred kilometres over land, beyond which signal levels rapidly decrease. Therefore, daytime operating tends to be 'ragchewing' contacts with fairly local stations and datacomms tests, although when the signal path is mostly over the sea, ranges of 600 - 700 km have regularly been achieved in full daylight.

At night, sky-wave propagation greatly improves signal levels over longer paths, enabling potentially world-wide DX operation. A characteristic of MF sky-wave propagation is deep fading that often results in strong signals disappearing into the noise every few minutes, returning a few minutes later. This makes it difficult to increase range simply by using extremely slow QRSS as is

CHAPTER 9: OPERATING PRACTICE

Frequency (kHz)	Callsign / ident	Location
474	BIA	Rzeszow, Poland
474.5	SA	Darlowo Poland
475	RP	Senica, Slovakia
480	VIB	Viterbo, Italy

Note that these are MCW (AM) carrier frequencies; the Morse ident appears as CW approximately 1kHz either side of the frequencies shown.

Table 9.1: Some NDBs commonly received in the UK within the 472kHz amateur band

normal on 136kHz. The 472kHz band is more suited to faster modes such as CW, Opera4, WSPR2 and QRSS3.

Activity levels for random two-way contacts tend to peak at the weekends, or when some special event occurs, such as an expedition station operating from a portable location. At other times, much of the activity is from beacon-style modes, for instance WSPR.

It is unlikely that permanent amateur radio beacons will be licensed for this band, but several aircraft beacons (NDBs) operate in Europe and these provide useful sky-wave propagation indicators. Note that although NDBs are frequently active for long periods, they do not necessarily run all of the time, so the absence of a signal is not in itself an indicator of poor propagation. Some of the NDBs most often heard in the UK are shown in **Table 9.1** and a comprehensive list can be found at [4]. Note that it is a condition of the UK Notice of Variation that amateurs must not cause harmful interference to non-amateur stations on this band, and this includes these beacons.

As on 136kHz, modes requiring a bandwidth greater than about 200Hz are not popular due to the narrow width of the allocation. The most obvious mode to use is CW. Also popular are computer-based modes such as QRSS/DFCW, Opera and WSPR which are capable of monitoring several stations simultaneously and offer a signal-noise improvement on CW, albeit at a sacrifice in time. There is much more on modes in a later chapter.

Bandplanning

No formal band plan currently exists for 472kHz, but **Table 9.2** shows the frequencies commonly in use for each mode at the time this book was written. As more and more countries allow access to 472kHz, and new digital modes are devised, the usage of the band is likely to change, For the very latest situation, see the RSGB LF Group [1].

CW	Often around 472.5kHz, though can be heard all over the lower half of the band.
WSPR	Set dial to 474.2kHz USB (for signals between 475.6 and 475.8kHz).
ROS	Set dial to 476kHz USB
QRSS	Around 476.175kHz. Also around 478.900kHz.
WSJT-X	Set dial to 477.0kHz USB (for signals between 478.0 and 478.5kHz) [other frequencies may be used]
Opera	Set dial to 477.0 kHz USB (for signals between 478.5 and 478.8kHz).

Table 9.2: Activity by mode on the 472kHz band (subject to change)

LF TODAY

G3LDO's 136kHz station. On the right is an ex- Decca system 1kW transmitter, with DDS driver, centre. Far right is the primary tuning and matching coil and variometer; the main loading coil is located outside the shack. The transmitter is remotely controlled from a shack in the house where the main receiver (TS-850) is located

Operating on 136kHz

There is occasional CW operation on the 136kHz band, but ranges are limited to a few hundred kilometres, so most people use more effective computer-based modes, such as QRSS. Much activity involves beacon modes, such as Opera and WSPR. See the Modes chapter for a discussion of relative features and requirements. Random QSO activity is mainly at weekends and evenings, especially during the winter months. Intercontinental activity may take place overnight.

Almost all European countries (including all of Russia) have an allocation at 136kHz, most with a 1W radiated power limit, and several UK operators have had two-way contacts with over 20 countries. Some experimental licences have also been issued in the USA, Australia and New Zealand, though not all are still current. There is also regular activity from Canada and Japan. At the time of writing, US amateurs are applying for general use of this band. Additionally, amateur LF transmissions have been received by enthusiasts in many parts of the world.

Bandplanning

The official IARU/RSGB bandplan says:

"135.8 - 137.8kHz; Necessary bandwidth 200Hz; UK usage CW, QRSS and Narrowband Digital Modes."

In other words, apart from specifying that transmissions should have a narrow bandwidth, there is no (UK) nationally or internationally laid down bandplan.

That does not mean there is a free-for-all, but it does mean the band users can decide amongst themselves where to operate for best results. For instance, the frequencies used for intercontinental QRSS have been changed from time to time by mutual agreement to avoid interference from commercial stations.

It is, nevertheless, important to know where particular modes can be heard, and to separate incompatible modes, so as with the other amateur bands 136kHz has a voluntary activity plan. **Table 9.3** shows how the band is divided into modes as well as certain specialist activities. Adhering to the plan gives the best

Table 9.3: Activity on the 136kHz band (subject to change)

135.700 - 136.000	Station tests
136.000 - 137.000	CW (avoiding the Intercontinental slot at 136.172kHz)
136.169 - 136-175	Intercontinental QRSS (Europe transmitting)
137.400 - 137.600	WSPR2 (set dial to 136.000kHz USB)
137.600 - 137.625	WSPR15 (set dial to 136.000kHz USB)
137.500 - 137.600	Opera32 (set dial to 136.000kHz USB)
137.600 - 137.700	Opera8 (set dial to 136.000kHz USB)
137.660 - 137.740	QRSS3 and QRSS10
137.774 - 137.780	Intercontinental QRSS (N America and Russia transmitting)

likelihood of successful contacts and minimises the risk of causing annoyance to other stations.

It can be seen that normal CW is in the centre of the band, data and experimental modes towards the top and very slow CW (QRSS) right at the top. The bottom extremity is used for testing. DX working is often split-frequency to enable intercontinental contacts where local transmissions would cause undue interference to the extremely weak DX signals.

QRM and QRN

After the inefficiency of practical antennas, the next limiting factor to success on the LF and MF bands is noise on receive. Noise can take several forms, and divide between man-made and natural noises, and again between local and distant. Some noise is unavoidable but others can be eliminated, or at least reduced to a tolerable level. The 136kHz and 472kHz bands are subject to higher noise levels than amateur allocations at HF and above. Low frequency bands are also shared with commercial stations. Interference in this frequency range has a number of origins as follows.

Natural band noise

Under quiet daytime band conditions, a low level background hiss is present, similar to the band noise in the HF range. But much of the time, the dominant natural noise source is a crackling static (QRN) produced by distant thunderstorms. In general, thunderstorm QRN levels are higher on 136kHz than 472kHz. QRN on both bands tends to be much stronger at night and during the summer. Static will also often be present during the daytime on 136kHz and, if there is nearby thunderstorm activity, also on 472kHz.

Usually it is not very effective to use a directional antenna to reduce QRN, because thunderstorm activity is normally present over wide areas, and the antenna only provides a deep null over a narrow range of angles.

Receiver noise blankers have been used to reduce the effects of QRN, with rather mixed results.

Non-amateur stations

Both the 136kHz and 472kHz bands are secondary allocations, shared with other spectrum users.

LF TODAY

The solid lines are Loran sidebands. In between there is a weak QRSS transmission. Care should be taken to avoid selecting a QRSS transmit frequency that coincides with a Loran line

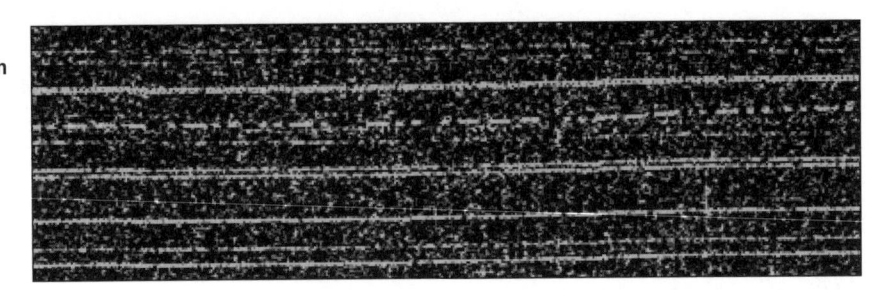

As can be seen from the plot shown in chapter 1, the 136kHz band has high power utility stations inside and outside the band. In the UK, the main interference comes from the sidebands of German station DFC39, just above the band on 138.830kHz, when it transmits bursts of data every few seconds. These bursts are usually a second or so long, but can last several seconds. A station in Hungary, HGA, operates just below the band on 135.43kHz, also has databursts, but these are often shorter and cause less of a problem in the UK. A Greek RTTY station, SXV, operates on 135.75kHz which is outside the most popular part of the band and is narrowband, so it does not present a practical problem. From Halifax, Nova Scotia, CFH transmits occasionally on 137.000kHz, and this can create difficulties for North American stations.

A widespread source of interference in much of the UK, Europe and the northern hemisphere are the sidebands of the Loran C navigation beacon system [5]. Although these pulsed signals are nominally confined to a range of 90 - 110kHz, sufficient leakage occurs outside this range for the rhythmic chattering 'galloping horses' sound to be the dominant 136kHz band noise at some locations.

The radio spectrum around 472kHz has much lower occupancy, so is relatively free from these kinds of interference, with the exception of the low power NDBs mention above. Spurious emissions and blocking effects on 472kHz can result from Long Wave and Medium Wave broadcasting stations, particularly if they are located close to the receiving station.

These types of unwanted signal can be effectively suppressed using the directional nulls of a loop or ferrite rod antenna, provided they are in a different direction to the wanted signal, though often only one interference source can be nulled at any one time. See the chapter on receive antennas for more on this subject.

Local man-made QRM

In this frequency range, most local sources of interference are associated with the mains wiring, and electrical or electronic appliances connected to it. Current European EMC regulations do not require testing of mains noise levels below 150kHz, and permit relatively high noise levels in the MF frequency range. It is usually found that appliances generate maximum mains noise levels in the LF/MF range. This sort of noise can propagate considerable distances along the mains wiring, via which it is coupled to the antenna.

Other potential sources of local QRM are PCs, monitors and associated equipment, TV sets, fluorescent lights, telecomms equipment and associated cables. Local QRM can sometimes be eliminated by finding and removing the source of interference. If this is not possible, major improvements can often be achieved by using a separate receiving antenna positioned to minimise noise pick-up.

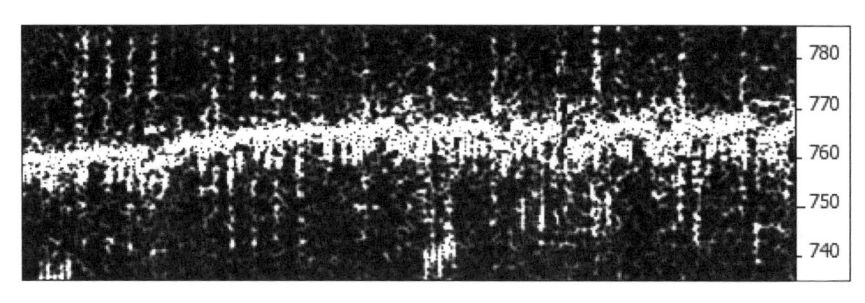

Spectrogram of interfering noise - probably from a switch-mode power supply. The vertical lines are databursts from DFC39

Identifying local interference sources

Many LF/MF operators are affected to some extent by local man-made QRM. A good first step in combating such interference is to identify likely sources from the nature of the noise being received. Many noise sources sound very similar when heard through a narrow CW filter, so try listening using SSB bandwidth; setting the receiver close to the bottom of the band in USB mode, or near the top of the band in LSB mode will make much of the amateur band audible.

A rough 50Hz buzzing noise covering a wide bandwidth typically is due to harmonics of the mains frequency. These may be generated by rectifiers, triac dimmers, motor speed controls and various types of lighting, or by high voltage overhead lines. Almost any mains-powered device containing a rectifier or other electronics can produce this kind of noise.

Drifty carriers with rough 50Hz modulation sidebands, often spreading over several kilohertz, can be caused by switched-mode power supplies. The carriers occur at the switching frequency or harmonics, and the frequency varies with temperature and loading, so often these noise sources drift in and out of the band over time, and are subject to abrupt frequency shifts. Switch-mode power supplies are increasingly used in all sorts of mains-powered appliances, even small plug-in chargers for mobile phones, etc, and are also used in energy-saving compact fluorescent light-bulbs, so are common interference sources (see picture).

Noise-like 'hash' sounds, often with a distinctive 'note', can be generated by TV sets and video monitors, and many digital devices such as PCs and their peripherals, computer games and data networking equipment. Occasionally, if the receiver is very close to the noise generating device, noise can be directly induced in the receiver circuits, but connecting cables are more likely to be the source of coupling between noise source and receiver.

A portable receiver with a directional antenna, such as a ferrite rod or a loop, can be used to try to locate the source of noise. However results can be confused by the noise being propagated by mains and other wiring.

If the noise source is equipment within your own house, it can usually be switched off while operating, or perhaps be replaced by a less noisy substitute. Switching off likely appliances while monitoring the noise level can identify the source. Often, it will be necessary to unplug the appliance to eliminate the noise, since many appliances have a 'standby' mode in which they can generate noise even when not actually operating; it is not unusual for the noise level to be higher in standby than when operating. Switching the mains supply off at the consumer unit is the most certain way of determining if noise sources are inside your home; this obviously requires a battery-operated receiver, and be prepared to have to reset timers, alarms etc!

LF TODAY

If the noise is coming from your neighbour's house, diplomacy is called for. Advice on tackling this sort of problem can be obtained from the RSGB [6].

Noise reducing antennas
The chapter on Receive Antennas describes how to use a separate receive antenna to reduce interference. In brief, a receive antenna should be located away from noise sources, may be directional, or may be used in conjunction with another antenna to cancel out a particular local problem.

Earthing
Some electrical noise can be induced into your receiver through earth paths. Often noise can be reduced by separating the 'radio earth' (see the chapters on Antennas) from the mains 'safety' earth. Please read the safety notice in the Antennas chapter about Protective Multiple Earthing before doing this.

Sometimes, noise pick-up can be made worse by earth loops between station equipment (eg radio and computer), and again this may be reduced by using separate earths at both ends of a connection, or even earthing just one end.

Unfortunately there are as many solutions as there are possible noise problems, so some experimentation is needed to achieve the lowest noise pick-up.

Receiving on loop or ferrite rod antennas requires no earth return, unlike a Marconi-style antenna. However, your equipment is likely to be connected to mains electricity, so there is still scope for noise to be introduced. Experimenting with an external 'radio earth' may well help if problems arise.

Remote receiving, grabbing and reporting

Thanks to high speed Internet access, it is easier than ever before to find out how your station is performing, even if there are no other human beings transmitting on the band. Automatic receivers and reporting tools make light work of experimentation, propagation research and activity monitoring.

Like many other amateur bands used by experimenters, 136 and 472kHz do not have large numbers of operators able to report on your signals 24 hours a day, seven days a week. Furthermore, these lower bands have their best DX propagation during the hours of darkness when many operators are asleep. As a consequence, much use is made of grabbers and remote reporting.

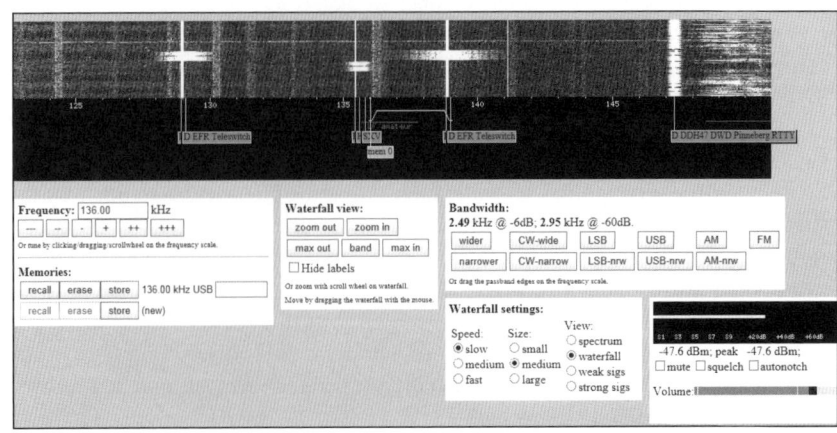

The remotely tuneable web receiver at the University of Twente. The 136kHz band can be seen in the centre of the waterfall display, together with utility stations HGA and DCF sending their datagrams

CHAPTER 9: OPERATING PRACTICE

> **G4WGT** (*http://myweb.tiscali.co.uk/wgtaylor/grabber2.html*): Grabbers for 73kHz, 136kHz and 472kHz, with a gallery of archived screen grabs.
>
> **DF6NM** (*http://www.df6nm.de/grabber/Grabber.htm*): Directional spectrograms, showing strength and direction of signals received in central Europe. QRSS3 and QRSS60 catered for.
>
> **YO/4X1RF** (*http://qsl.net/4x1rf/yo/lfgrabber.htm*): Bucharest. Several 136kHz grabbers at different speeds.
>
> **4X1RF** (*http://qsl.net/4x1rf/lf/lfgrabber.htm*): Haifa. Several 136kHz grabbers at different speeds.
>
> **TF3HZ** (*http://simnet.is/halldorgudmunds/TF3HZ_LFgrabber*): Useful mid-Atlantic grabbers for 136kHz and 472kHz.

Table 10.4: Selection of screen grabbers on 136kHz and 472kHz

Remote receivers

There are several Software Defined Radio (SDR) receivers that can be operated remotely via the worldwide web by many stations simultaneously. Some of these cover the lower bands, and can be used to check that your station is getting out, and for comparitive tests on antennas. A particularly useful one for European stations is the wide-band SDR at the University of Twente in the Netherlands which can be found at [7].

Grabbers

Using 'visual' modes such as QRSS and DFCW, a large number of signals can be monitored simultaneously, provided they fall within the bandwidth of the spectrogram display and the receiver filters. This has led several LF/MF enthusiasts to implement 'screen grabbers', where the spectrogram display is periodically uploaded to a web page, typically once every few minutes. Since receivers and antennas will be required for other purposes sometimes, most screen grabbers are not permanently operating, but are activated on request, or when there is interesting activity scheduled to occur. The displayed bandwidth, and rate of screen updates, is varied to suit the transmission modes expected. Up-to date information on operational screen grabbers can be obtained via the RSGB LF Group [1], or on individual operator's web sites. Some well-known screen grabbers are shown in **Table 10.4**. Some of the pictures of QRSS in this book were taken from grabbers.

Signal reporters

In the past, reports on your signals would come from short-wave listeners (SWLs) in the form of letters or QSL cards received days, weeks or even longer after your transmission. Then it became possible to report signals almost in real time using the DXCluster system, first by packet radio, then via the Internet. Nowadays many reporting systems are fully automatic and do not need a human operator at all.

Many computer-based modes can pack several stations in a narrow frequency slot, well within an SSB bandwidth. The software for these modes can automatically upload what a station hears to a central database which can be accessed via a web page. The Opera mode will actually display the database on screen. The databases can usually be sorted by band, mode, date and callsign,

LF TODAY

Part of the 472kHz listing of the Reverse Beacon network, showing reports on EI6IZ's CW transmission

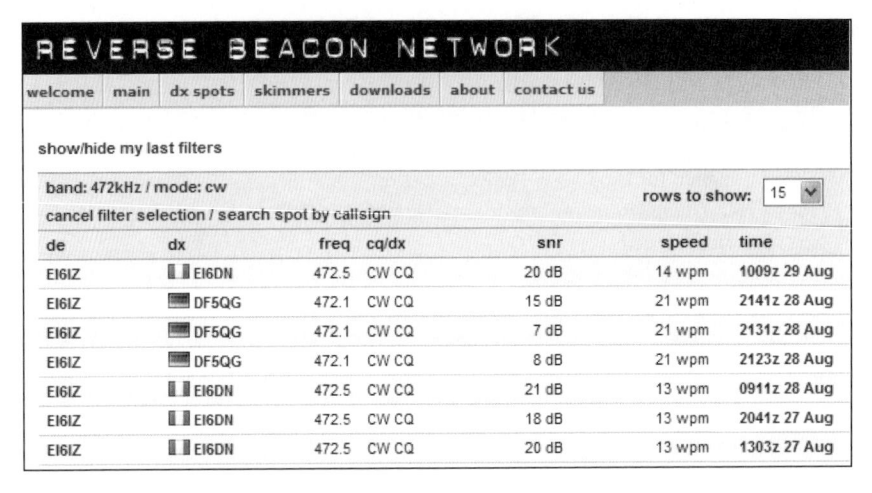

de	dx	freq	cq/dx	snr	speed	time
EI6IZ	EI6DN	472.5	CW CQ	20 dB	14 wpm	1009z 29 Aug
EI6IZ	DF5QG	472.1	CW CQ	15 dB	21 wpm	2141z 28 Aug
EI6IZ	DF5QG	472.1	CW CQ	7 dB	21 wpm	2131z 28 Aug
EI6IZ	DF5QG	472.1	CW CQ	8 dB	21 wpm	2123z 28 Aug
EI6IZ	EI6DN	472.5	CW CQ	21 dB	13 wpm	0911z 28 Aug
EI6IZ	EI6DN	472.5	CW CQ	18 dB	13 wpm	2041z 27 Aug
EI6IZ	EI6DN	472.5	CW CQ	20 dB	13 wpm	1303z 27 Aug

and will include a signal to noise ratio measurement. These automatic systems have the advantage that there are usually people who have left their receivers running, reporting on what they hear, and this is where non-transmitting stations can make a valuable contribution to the experiments of those who are transmitting. More on the various modes used on 136kHz and 472kHz, including pictures of the databases, can be found later in this book.

Reporting systems are to some extent mode dependent. For instance, for CW reports the Reverse Beacon Network [8] is useful. PSKreporter [9] covers many modes, including PSK and Opera, and displays real-time information on a world map. WSPRnet [10] displays a real-time database and map of WSPR stations together with reports on their beacons.

DX working

Like the 1.8MHz band, daytime propagation on 136 and 472kHz is restricted to ground-wave which limits maximum distance to a few hundred kilometres. At night, however, sky-wave propagation takes over, greatly increasing ranges. But this can also introduce noise and fading. A further disadvantage is that humans tend to sleep during much of the night, reducing the number of active stations and

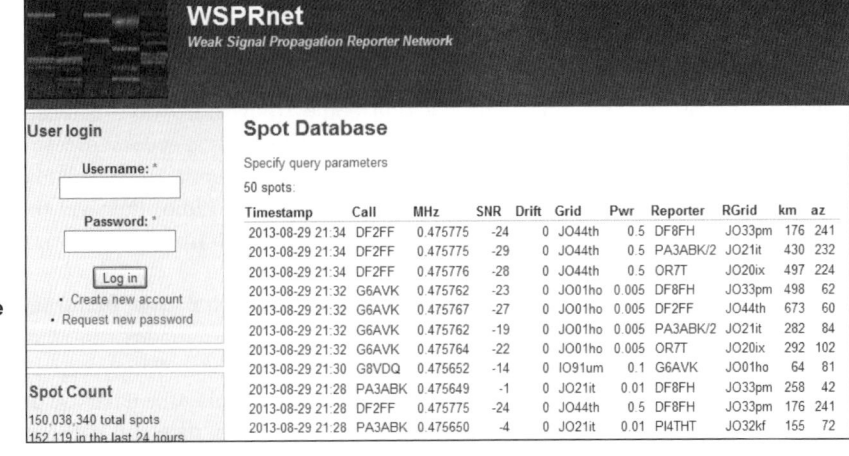

Part of the WSPRnet database display showing reports on 472kHz WSPR transmissions

Timestamp	Call	MHz	SNR	Drift	Grid	Pwr	Reporter	RGrid	km	az
2013-08-29 21:34	DF2FF	0.475775	-24	0	JO44th	0.5	DF8FH	JO33pm	176	241
2013-08-29 21:34	DF2FF	0.475775	-29	0	JO44th	0.5	PA3ABK/2	JO21it	430	232
2013-08-29 21:34	DF2FF	0.475776	-28	0	JO44th	0.5	OR7T	JO20ix	497	224
2013-08-29 21:32	G6AVK	0.475762	-23	0	JO01ho	0.005	DF8FH	JO33pm	498	62
2013-08-29 21:32	G6AVK	0.475767	-27	0	JO01ho	0.005	DF2FF	JO44th	673	60
2013-08-29 21:32	G6AVK	0.475762	-19	0	JO01ho	0.005	PA3ABK/2	JO21it	282	84
2013-08-29 21:32	G6AVK	0.475764	-22	0	JO01ho	0.005	OR7T	JO20ix	292	102
2013-08-29 21:30	G8VDQ	0.475652	-14	0	IO91um	0.1	G6AVK	JO01ho	64	81
2013-08-29 21:28	PA3ABK	0.475649	-1	0	JO21it	0.01	DF8FH	JO33pm	258	42
2013-08-29 21:28	DF2FF	0.475775	-24	0	JO44th	0.5	DF8FH	JO33pm	176	241
2013-08-29 21:28	PA3ABK	0.475650	-4	0	JO21it	0.01	PI4THT	JO32kf	155	72

Spot Count: 150,038,340 total spots; 152,119 in the last 24 hours

150

CHAPTER 9: OPERATING PRACTICE

often requiring a very late night on the part of one or both ends of a DX contact. All this will be familiar to the 1.8MHz DXer.

472kHz

At the time of writing, the 472kHz band is not yet available in North America (except to specially licensed experimental stations) or Russia, and this has so far limited the scope for long distance contacts from Europe. Some US stations using special experimental licences have been received in Europe, as has an Australian station. These have used Opera8 and WSPR2 beacon modes, but CW DX should be possible given the right conditions and well-equipped stations.

Intercontinental contacts are likely to become much more common when more countries adopt this band.

136kHz

On the 136kHz band, CW contacts at LF are restricted to a few hundred kilometres, but by using sophisticated techniques and a well-engineered station it is possible to be received at distances of 6000km or even more. The use of the word 'received' was deliberate as there are few transmitting amateurs outside Europe and Russia. Nevertheless, there are stations in north America and elsewhere who are equipped to receive on LF and are prepared to monitor for 136kHz transmissions. Some of the USA have special licences to make test transmissions, but not to make contacts.

There have been contacts between western Europe and Canadian amateurs and also with an expedition in Asiatic Russia, as well as trans-Pacific contacts, so two-way DX is indeed possible. But until US stations have general permission to transmit, most activity concentrates on one-way transmissions.

Intercontinental contacts are beyond the range of ground-wave so all propagation is via the ionosphere (see the chapter on propagation). Unfortunately, with the power levels available to amateurs - a maximum of one watt ERP in most cases - the day-to-day sky wave will not support reliable propagation of amateur-level signals beyond 2000km or so. As with much amateur radio communication, the trick is to wait for enhanced conditions and try to be on the air when they occur.

LF conditions for the necessary multi-hop path vary according to solar activity and darkness levels over the length of the path. Signal levels can vary considerably from day to day and from hour to hour as shown by the graphs overleaf. For anyone but the best-equipped stations to stand a chance of working DX, conditions must be close to the peak of W3EEE's 'best' graph. As can be seen, these last only a couple of hours on a good night and often much less.

So why is this a problem? As detailed in the Modes chapter, signal to noise ratio can be greatly improved by reducing the received (and transmitted) bandwidth. The downside of a reduced bandwidth is a slower rate of information transfer (Morse speed or data rate). Using a typical DX mode on 136kHz, it will take anything between 30 minutes and several hours for one station to transmit basic QSO information.

For example, in order to achieve ranges of 3000+km using QRSS it is necessary to use dot lengths of 30, 60 or 120 seconds, with the slowest rate having the best chance of being visible. But when using QRSS120, a callsign takes

LF TODAY

W3EEE's real time plot of the German commercial station DCF39 as received at his Pennsylvania site. The fat, lower line is the background noise level - note that the ionosphere propagates noise, too. The thin line above it is DCF39. The top graph shows a typical day with a short peak between 0130 and 0300UTC. The lower graph shows one of the best ever periods of transatlantic propagation with excellent conditions between 0000 and 0600UTC

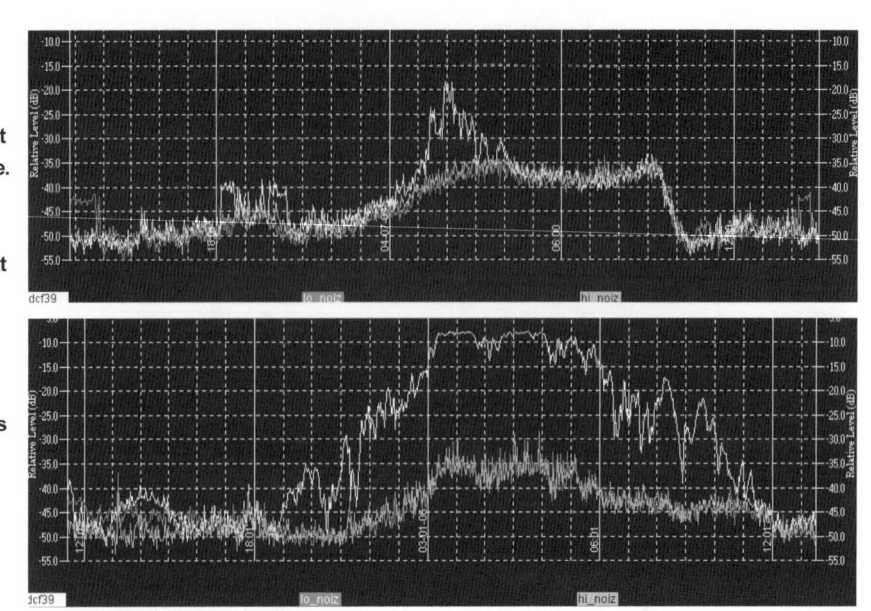

about two hours to send, and unless it starts exactly when the conditions start to peak it is unlikely that an entire call will be received. And that's just a single callsign; a two-way contact is much more of a problem - in fact, one of the earliest successful trans-Atlantic two-ways was spread over several nights in order to have enough total time to complete the contact.

So the difficulty with DX communication at 136kHz is the time period of the good conditions versus the speed necessary to get an adequate signal to noise ratio. Four strategies have been devised to try to overcome this problem.

Firstly, it is important to be aware of when propagation conditions are likely to be good - or more correctly when conditions will be so poor it isn't worth bothering. It is possible to get an idea of conditions from checking the day-to-day reports of solar storms (see the chapter on propagation).

Next, screen grabbers and other remote monitoring systems as described above can be used to check conditions in real-time provided there are stations active. Activity is often from beacon modes such as Opera32 and WSPR.

Listening on 137.000kHz may reveal the high powered Canadian naval station CFH. CFH is often off the air so its absence is not a useful guide, but if it is 'S'9 in Europe conditions can be expected to be suitable for amateur DX. Similarly, stations outside Europe can check the strength of utility stations DCF and HGA.

Dual frequency CW (DFCW - see the Modes chapter) can be used to reduce the transmission time without reducing the signal to noise ratio. Because DFCW occupies a wider frequency 'slot' than QRSS, more care than usual must be taken to avoid clashing with other operators.

Lastly, choosing the right mode to make full use of the relatively short periods of very good propagation may help. Some modes, such as WOLF, Opera32 and WSPR15 are designed specifically for LF use.

Split frequency is almost always used for intercontinental QRSS working. Europeans use a slot which is currently close to 136.170kHz, but this may change depending on noise levels at the receive end.

CHAPTER 9: OPERATING PRACTICE

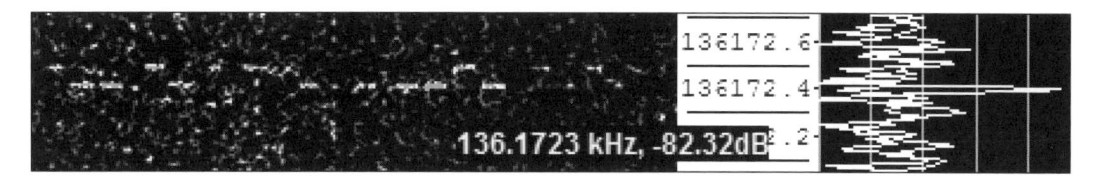

The DFCW transmission from high powered station DK7FC received by Australian SWL Edgar J Twining in April 2013. The 'F' of his callsign can clearly be seen, and with a little care the rest of the callsign can just be made out

North Americans, and often eastern Russians, use a slot around 137.775kHz. Because a typical receive spectrogram screen is just 3Hz wide when receiving QRSS60 and half of this for QRSS120, all stations must operate within a slot no more than a couple of Hertz wide. It is important to check activity to avoid a clash of frequency.

Most DX activity is well publicised [1] and most of the two-way contacts have been scheduled, although a couple have been random. The exchange of information is kept to the absolute minimum of callsign, report and 'roger'.

Operating away from home

Many stations have conducted expeditions using LF. Some have shacks set up at alternative locations, others have operated out of doors, whilst a few have had the privilege of a few hours' use of a massive commercial mast. In all cases, the effort involved has been worthwhile in terms of the fun of overcoming difficulties and in terms of the activity it has generated from other band users keen for a contact.

By far the easiest of these is the permanently set up alternative shack, although of course establishing it in the first place entails all of the complexity of any LF installation. This includes getting a good earth connection, erecting or locating high antenna supports and tuning the system.

Outdoor LF activity is great fun, although quite difficult. Your transmitter must be capable of running from a car battery (keep the engine running or the battery may flatten rather rapidly), or from a generator. A petrol driven generator providing 240VAC can be hired cheaply, but beware that it may also generate LF noise. The antenna support can be a tree - be careful to use good insulators - a high cliff or a building. Kites or balloons are convenient as they can be used anywhere, although they can be very dependent on the right weather. It is also important to master the flying techniques before using one for RF. More information on kites and balloons can be found in the chapter on Antennas and Matching.

Safety is a vital issue, especially if members of the public are close by. Your activity may attract interested locals, even the police or coastguard, so keep your licence handy and perhaps a leaflet on amateur radio to save endless explanations. Tales of operating from alternative and portable locations can be found at [11].

Several amateurs, usually in groups, have been able to 'borrow' a tall mast - sometimes 100m or more high - from a commercial operator. This is sometimes between the decommissioning of the station and the demolition of the mast. This must never be done without the permission of the mast owner as there are safety (and corresponding financial) issues involved. It may seem that this is an easy way to operate 'portable', as there is often mains power and a building to operate from, as well as the likelihood of the biggest antenna you will ever use. However, if the mast is earthed, it may be difficult to use it as an antenna. You

LF TODAY

This 100m high Decca Navigation mast was available for amateur use for a weekend. The mast was insulated which made the antenna simple. On the right is G3KAU and G3YXM in the equipment hut. Note the antenna feed-through in the window, with the lightning arrester below. More information can be found at [12]

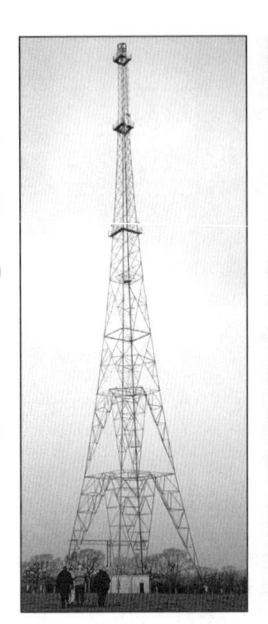

may be lucky and find a mast intended for LF work, or a centre-fed T antenna for MF use. A mast with its own staircase may make it easier to erect a wire, but always get permission to climb and never climb the outside of a lattice tower.

Although impressive signals can be radiated in this way, high received noise levels at the expedition site are a frequent problem; a big antenna often also picks up an impressive level of noise! The result is disappointingly few contacts. If you are planning an expedition, it is a good idea to have available alternative receiving antennas of the types described in an earlier chapter, along with some long lengths of coax feeder. Strong out-of band signals may also play havoc with equipment used for tuning up. If at all possible, try receiving at the expedition site before transmission begins, so that noise problems can be investigated and remedied.

References

[1] http://uk.groups.yahoo.com/group/rsgb_lf_group/
[2] *RSGB Yearbook*, published annually by the RSGB.
[3] Calculator for locators. http://www.qsl.net/dl3bak/qrb/en/frame.htm
[4] http://www.classaxe.com/dx/ndb/reu/
[5] Loran-C: http://en.wikipedia.org/wiki/LORAN
[6] RSGB EMC Committee: http://rsgb.org/main/technical/emc/
[7] http://websdr.ewi.utwente.nl:8901/
[8] http://www.reversebeacon.net/
[9] http://pskreporter.info/pskmap.html
[10] http://wsprnet.org/drupal/
[11] G3YXM's LF expeditions. http://www.wireless.org.uk/136gm.htm and http://www.wireless.org.uk/features.htm
[12] The Puckeridge expedition. http://homepage.ntlworld.com/mike.dennison/index/lf/decca/index.htm

10

Modes used on 136 and 472kHz

In this chapter:

- On-off keying modes
- Multi-frequency modes
- Modes requiring a linear tx
- Better synchronisation
- Modes compared
- Interfacing with a computer

ALTHOUGH CW CAN BE (and is) used on the low bands, it is possible to increase ranges greatly by selecting the most appropriate mode for the propagation conditions. Furthermore, relatively low activity levels mean that personal beacon stations and remote reporting play a large part in helping people experiment with their equipment, antennas and propagation. Modes, too, are the subject of much experimentation, and new modes appear quite frequently. Broadly speaking, all modes used on 136kHz and 472kHz divide into three categories, the choice of which to use being dictated by your transmitter architecture.

On-off keying modes

These are still popular and involve the simplest equipment. A simple keyed oscillator and a Class D/E power amplifier are all that is required. Despite the simplicity of on-off keying, high communication efficiencies can be achieved.

Hand-keyed CW

Popular for relatively short-range chatting, CW is sent at between 10 and 20WPM. Nowadays, CW is rare on 136kHz because of high noise levels and low signal strengths. However, it is popular on 472kHz where the rapid fading on sky-wave signals creates problems for slower modes. CW beacons are unusual, but some stations using other modes will identify in CW from time to time. QSO practice is much the same as on the HF bands, though a QTH locator may be exchanged so that distance can be calculated easily.

Slow CW - QRSS/DFCW

QRSS uses Morse code sent at an extremely slow speed. There's no need to learn the code as any received characters could be looked up in a table by the receiving station. QRSS originated when UK amateurs first started operating at LF, on the now defunct 73kHz band. It quickly became apparent that there were

Reception of the first ever extremely slow CW amateur signals by G3PLX. It took nearly three hours to send, but G4JNT's 1mW ERP 73kHz signal travelled 400km to set a new distance record

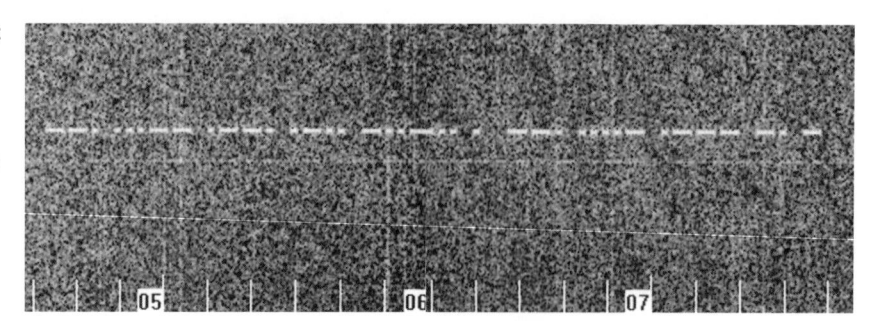

severe limits on the range of CW, especially for those whose small gardens obliged them to run extremely low radiated power.

Since the signal level could not be improved, received noise was the limiting factor. Noise is directly proportional to bandwidth so ways were sought to reduce the bandwidth by a significant amount, well below that required for normal CW, whilst still retaining the ability to transmit information.

In 1997, Peter Martinez, G3PLX, built a system that would display on a computer screen the processed audio output from his 73kHz receiver. The system used a hardware digital signal processing (DSP) unit together with Fast Fourier Transform (FFT) performed in software by his computer. The screen showed the level of signal plus noise in a bandwidth of just 0.025Hz (that's 25 millihertz!). When just noise was being received, the screen showed apparently random dots, but whilst a signal was present a line appeared. This enabled the reception of G4JNT's callsign in Morse over the previously unheard of distance of 393km. The transmission power was just one milliwatt ERP and the signal was completely inaudible at the receive end.

So far, so good, but the transmitted bandwidth of a Morse signal is directly proportional to the keying speed. So what speed can be received in a bandwidth of 25mHz? It is just 0.15 words per minute (WPM), requiring dots 80 seconds long and dashes of 240s. In fact, "G3PLX de G4JNT" took all of three hours to send. Plainly, this type of transmission is not for decoding by ear, nor is there much time for ragchewing.

Table 10.1 shows the relationship between speed, bandwidth and signal to noise ratio (SNR). It can be seen that for marginal paths (ie most paths on 136kHz and some DX paths on 472kHz) QRSS can provide a dramatic improvement in signal to noise ratio but at the expense of time.

Table 10.1: Effect of reducing transmission speed

Speed (mode)	Optimum bandwidth (Hz)	SNR vs 12WPM (dB)
12WPM (CW)	10	0
8WPM (CW)	6.67	+1.8
4WPM (CW)	3.33	+4.8
1 sec/dot (QRSS1)	1	+10
3 sec/dot (QRSS3)	0.33	+14.8
10 sec/dot (QRSS10)	0.1	+20
60 sec/dot (QRSS60)	0.0165	+27.8
120 sec/dot (QRSS120)	0.008	+30.8

CHAPTER 10: MODES USED ON 136kHz AND 472kHz

Some time after G3PLX's record-breaking test, freely available software originally written to analyse birdsong was found to perform a similar function to G3PLX's microprocessor, but using a computer's soundcard. This was used to receive DA0LF's experimental 136kHz transmissions - the first signals heard from outside the UK.

Nowadays, several amateurs have produced soundcard software specifically for LF use, the most popular being the simple *Argo* and the complex but versatile *Spectrum Laboratory* (*SpecLab*). These are capable of displaying Morse with dot lengths from less than one second to 1200 seconds. Further information on this software can be found in Appendix 2.

It is not the aim of this book to go into technical detail on this subject, but a very useful explanation can be found at [1].

Three hours to send a couple of callsigns may not sound like much fun, so in practice a dot length of three seconds ('QRSS3') is the default QSO mode. This allows a basic 'rubber-stamp' two-way contact to take place in a little over half an hour. Slower speeds are used, but usually only for intercontinental DX on 136kHz (see chapter on Operating).

QRSS reception: The advice to listen first is even more important with QRSS than other modes. In fact, it is look first, as QRSS is a visual mode. The centres of activity for QRSS3 (three second dot length) are 137.700kHz, 476.175kHz and 478.900kHz (subject to change). The spectrogram screen will display stations within 30-40Hz of this frequency. You may be able to hear QRSS transmissions, recognisable as three-second dots and nine-second dashes. Beware that the Opera mode (see below) sounds and looks similar to QRSS, but is not Morse code so will be impossible to read by eye or by ear.

To read QRSS you will need a PC-type computer with a soundcard. An old PC will work, but the more sophisticated programs may require a more modern, faster machine. Next, download the software. *Argo* is recommended to the beginner as it is the simplest to use. See Appendix 2 for more details.

Connect the audio output of your LF receiver to the input of the computer's sound card (see below for advice on the best way to do this). Run the *Argo* program and a scrolling display should appear. Signal strength is shown by brightness, and frequency by position on the screen. It is not the intention of this book to provide a detailed instruction manual for software - they all have Help files - just play with the controls until you have the hang of it.

If nothing is seen on the *Argo* screen, check the audio connections and if necessary switch the receiver to a strong signal such as the BBC on 198kHz. Note that the correct sound card audio input may first have to be enabled via the Windows "Volume Control" accessory, in order for the receiver output to be displayed. It is also advisable to disable any unused audio inputs. Volume Control can be accessed by right clicking with the mouse on the Windows task bar loudspeaker icon, and selecting "recording devices" or similar.

Once set up, the screen should look similar to one of the two views in **Fig 10.2**. On the left is a 'waterfall', scrolling vertically. The vertical line is a constant carrier; the horizontal lines are static crashes. The screen is 1.5kHz wide so it can be used to monitor most of the band, though the picture shows the effect of a 500Hz IF filter. The right hand picture shows the QRSS horizontal display monitoring the same carrier and the mode set to QRSS3 (three-second

LF TODAY

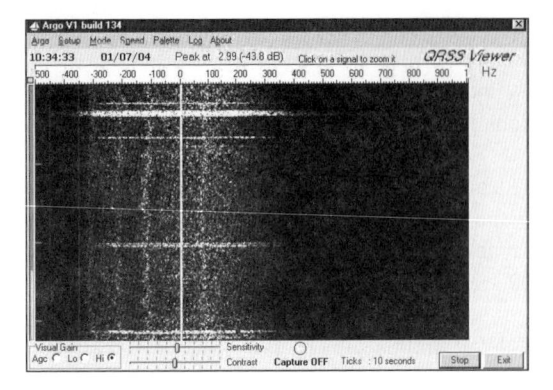

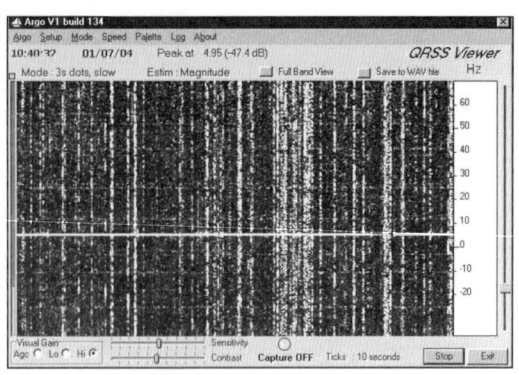

Fig 10.2: Two types of display on *Argo*: the vertical 'waterfall' (left) and the horizontal QRSS screen. The solid line is a constant carrier and the lines at 90 degrees to it are static crashes.

dots). The static crashes are now vertical and the total height of the screen is about 100Hz. This display is used for decoding the QRSS signals that will appear as dots and dashes horizontally across the screen. The resultant Morse code is deciphered by eye. There is no hurry as even QRSS3 takes about three minutes to display a callsign.

For best reception, the software and audio levels should be set so that noise forms a dim background on the screen. On the 136kHz band, it is likely that Loran-C sidebands (see later) will also be present - they can be seen as the faint horizontal lines in the right hand picture.

QRSS signals at different dot lengths can be received simultaneously but best reception is when the receive software is set to the correct QRSS mode. If the receiver is set too fast (wider bandwidth) the signal to noise ratio will be poorer and the Morse will appear 'wide' on the screen. If it is set too slow (narrower bandwidth), the signal will be blurred, often running the dots and dashes together to make it unreadable.

It is possible to save to disk an image of the screen, either by clicking on a button or automatically at regular intervals. Distant stations will appreciate being e-mailed a screen shot of their signals.

QRSS transmitting: Having seen and successfully decoded some QRSS signals, it is now time to try transmitting. Because of the long transmissions, it may be necessary to make sure the transmitter cooling is adequate. Also, this is the mode that will really test whether the antenna and tuning components are going to flash over. Be vigilant during your first few QRSS transmissions.

Although it is possible to send QRSS with a key and a stopwatch, the most straightforward, and most popular, is to use software and a small interface (such as the one shown later in this chapter) to connect to the PC. ON7YD's *QRS* program is designed to send all types of QRSS transmission and is freely available. The software is designed to run simultaneously with a spectrogram program on the same PC. More on this in Appendix 2.

Making a QRSS contact: Because of the time taken to send QRSS signals, a simplified QSO format has been devised. The 'de' between callsigns is not used and reports are in the form 'O' for perfectly readable, 'M' for readable with difficulty and 'T' for visible but not readable. Callsigns are rarely repeated and are abbreviated after the first full call. However, in common with standard amateur radio practice, a contact consists of an exchange of callsigns and reports and the final 'rogers'. A typical contact, lasting about 40 minutes in QRSS3, is shown

CHAPTER 10: MODES USED ON 136kHz AND 472kHz

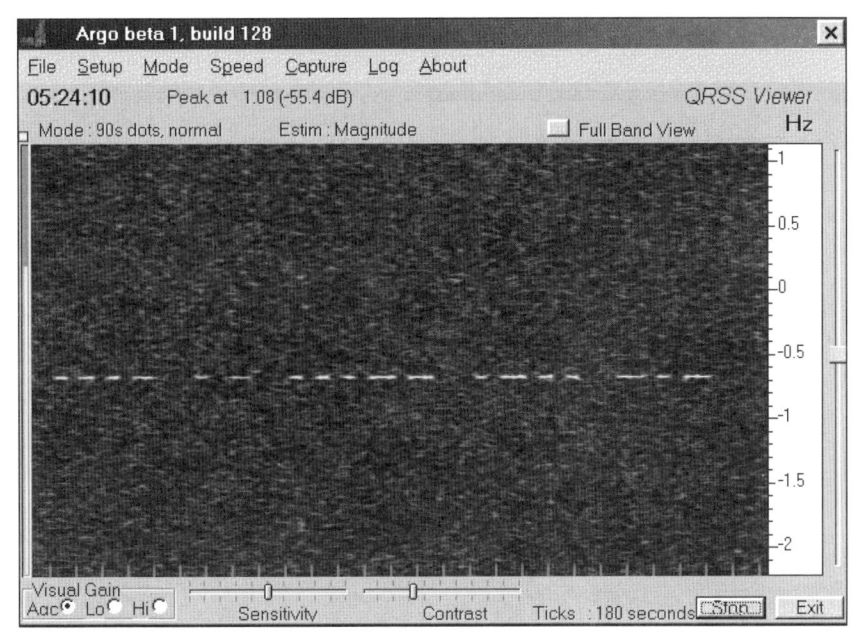

VA3LK received by CT1DRP using the Argo software set to receive a dot period of two minutes (QRSS120). Note that to reduce transmission time in this case, the dashes are just two dot-lengths

below (the locator is optional and would be omitted if it is already known or in marginal contacts).

```
CQ ON7YD K
ON7YD G3XDV K
XDV YD O O JO20IX K
YD XDV R M M IO91VT K
YD R TU SK
```

If someone hears just the end of this contact and wishes to call one of the stations, he may not have seen the full callsign. An abbreviated call may be used provided full calls are sent and received at some time during the contact. Eg:

```
YD DL3LP K
DL3LP ON7YD O O K
ON7YD LP R O O K
LP YD R TU SK
```

Alternatively a single question mark may be sent, meaning 'QRZ?'.

```
? DL3LP K
```

DFCW: Because even abbreviated QRSS is still time consuming, a variant has been devised to speed up the information transfer whilst still retaining the same bandwidth (and hence signal to noise ratio). DFCW (dual frequency CW) uses two closely spaced frequencies, the lower one for dots, the higher for dashes, and can be received using exactly the same software and settings as QRSS. Dot/dash frequency spacing is not critical but is typically 12 divided by the dot length in seconds (eg for QRSS3 it is about 4Hz and for QRSS120, 0.1Hz). The time savings are made by making the dashes the same length as dots, and having only a very short gap between each dot or dash (**Fig 10.3**), resulting in up to 50% time

LF TODAY

Fig 10.3: Dual-frequency CW (DFCW) uses the same length dots and dashes. Can you read this callsign? Tip: say 'dah' for each white bar on the top line and 'dit' for each on the bottom line

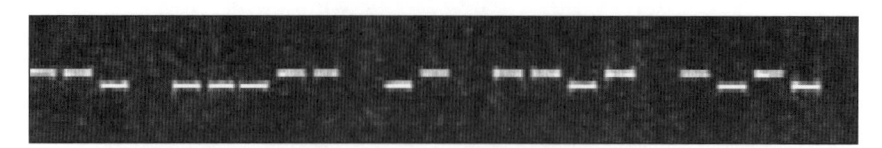

improvement. In all other respects it is used in exactly the same way as QRSS, and cross-mode contacts are common.

If DFCW has these advantages, why is it not used universally? Well, its main disadvantage is the need to switch between two accurate frequencies. It can also be slightly harder to decipher by eye than simple QRSS under noisy conditions. Nevertheless, it has been used successfully for intercontinental transmissions where the reduction in transmission time has a distinct advantage (see the chapter on Operating).

Strictly, DFCW should be included with the multi-frequency modes below. However, since only two frequencies are involved it may be possible to devise a frequency shifting mechanism (eg varicap diode) that is far less complex than the equipment required for other multifrequency modes such as WSPR.

Opera

Not to be confused with the internet browser of the same name, Opera is a mode that combines the effectiveness and simplicity of QRSS with the facilities available with many datamodes. Any type of CW transmitter may be used.

Reception of Opera by RX3DHR on the 136kHz band. The left hand list shows stations received by RX3DHR himself, whilst the right hand list is reported reception by others on the same band

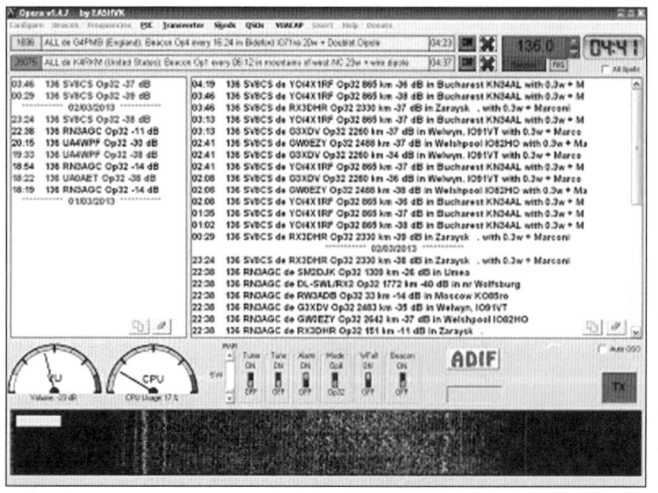

At the time of writing, Opera works only as a personal beacon (on the LF/MF bands), and no QSO mode is available. Nevertheless it is popular, especially on 136kHz, as a way to explore propagation effects, and to determine the effectiveness of a station.

Opera transmits only the station's callsign in short and long 'dots', but in a different code from Morse, and in a way that provides some resistance to fading and interference. This coding enables the receiver to decode the callsign automatically and display it on screen. The program is usually operated whilst connected to the internet (though this is not essential) and this allows a central database to calculate the distance between transmitter and receiver, and display it on the Opera screen, together with brief information about the transmitting station. Additionally, by accessing the web site *PSKreporter* [2], real-time information can be seen on a map, showing who is transmitting or receiving on Opera on each band.

The default Opera modes on 136kHz are Op8 which has a total transmission time of eight minutes and Op32 which lasts 32 minutes and can receive stations some 6dB weaker than the faster mode. On 472kHz, the default modes are Op4 and Op8, these faster modes being more suited to the sky-wave fading pattern on this band. The effectiveness of these modes can be compared approximately to QRSS as follows: Op4 ≅ QRSS3, Op8 ≅ QRSS10 and Op32 ≅ QRSS30.

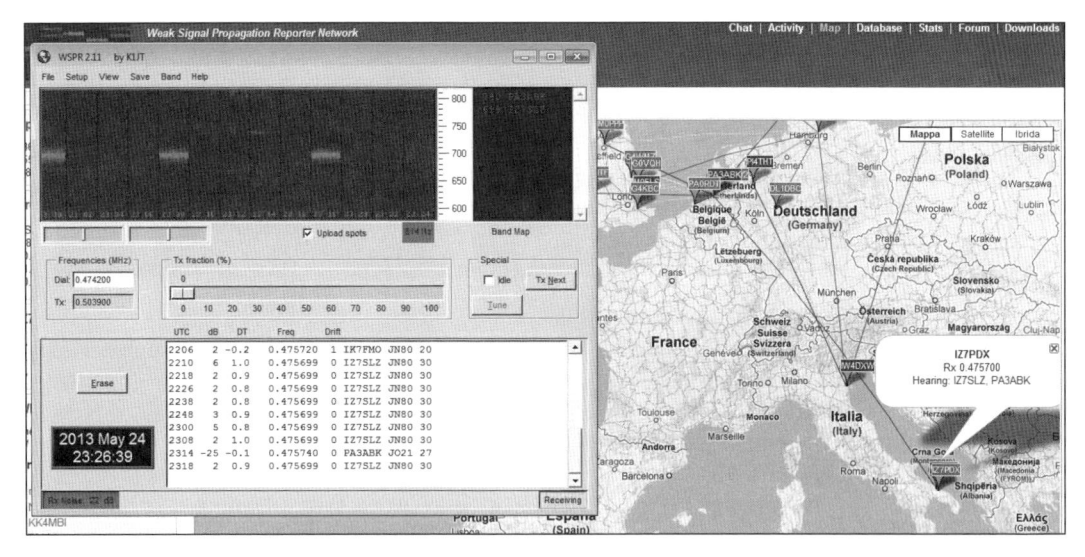

The software is available via a Yahoo group [3] and needs a fairly modern computer as the receive side is quite resource hungry. If in doubt, download it and check. Receive input is by connection to your sound card, and transmit output is a simple Tx on/off via an interface such as the one shown in this chapter.

WSPR in action. On the left is the communication screen showing the time/frequency display and the decodes; in the background is the activity map from WSRPnet.org [picture: forum post by IZ7PDX on hamradioweb.org]

Multi-frequency modes

If your transmitter is able to convert audio tones to frequencies on 136kHz or 472kHz (see the chapter on Generating a Signal), a number of modes are available in addition to those shown above, by using the appropriate software. Experiments have taken place using RTTY, ROS, WSJT-X and others, but the multi-frequency modes most commonly used on LF/MF are shown below:

WSPR

Pronounced "whisper", these initials stand for Weak Signal Propagation Reporter. WSPR was designed for probing potential propagation paths by using low power on the HF bands. WSPR signals convey a callsign, locator, and power level using a compressed data format with strong forward error correction and narrow-band 4-FSK modulation. It is a beacon-only mode with each transmission contained in a fixed two minute slot. Several stations can be decoded simultaneously. The software is available from [4].

WSPR is very popular for propagation experiments on the 472kHz band, and to a lesser extent on 136kHz. Low power MF operators have had impressive results with this mode. Stations with internet access can automatically upload their reception reports to a central database called WSPRnet [5], which includes a mapping facility.

An LF-only variant, WSPR-15, has a much slower data rate and uses a 15 minute slot. This enables weaker signals can be received and it has been used successfully over transatlantic paths on 136kHz.

161

SMT Hell transmits only one tone at a time, and occupies a bandwidth of less than 10Hz

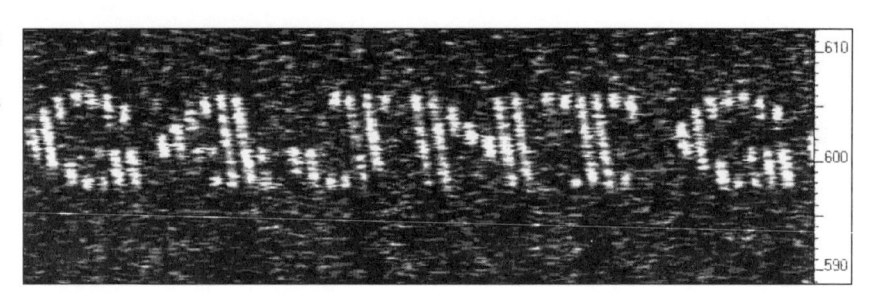

Sequential Multitone Hell

This strangely named mode is an electronic version of the mechanical Hellschreiber developed by Rudolph Hell in 1927. Hell is closer to fax than a conventional data mode in that each character is displayed as an image. Modern PC sound card based software reproduces the Hellschreiber machine output.

Narrow-band Hell modes have been developed specifically for use on the low bands. Unlike Opera and WSPR, Hell can be used for two-way contacts.

SMT-Hell is an MFSK mode, using several audio tones but transmitting only one at a time (which is why the text appears sloping), so a simple non-linear Class D/E transmitter can be used. Some Hell variants transmit several tones simultaneously, and these require a linear transmitter.

The software is available as a component of DL4YHF's *SpecLab* software [6]. The speed and bandwidth are compatible with QRSS and DFCW signals, and signals can be read alongside them, so mixed-mode contacts are possible.

Jason

As an extension of the slow CW technique, the writers of *Argo* have created a data mode called Jason which uses multiple tones to transmit data at a rate of about 2.5 seconds per character. It is a keyboard to keyboard mode with a bandwidth of about 6Hz. The 'send' output is either tones for input to an SSB transmitter, or serial data for a DDS. Jason is capable of very good performance but can be susceptible to interference from Loran sidebands. Further details can be found at [1 and 7] and the software is at [8].

Computer control of DDS

Multitone modes are often generated as audio signals and then used to drive an SSB transmitter transverted to LF/MF. Alternatively, data is used to control a DDS at the signal frequency. This is the method used by G0UPL's Ultimate QRSS kit, an inexpensive way to generate modes such as WSPR and SMT-Hell (see the chapter on Generating a Signal).

Modes requiring a linear transmitter

Although most low frequency operation uses modes suitable for high-efficiency class D/E power amplifiers, several data modes have been experimented with that need a linear amplifier. Beware that using a non-linear PA for these modes will result in a wide transmission that will make you *very* unpopular with other band users. Suitable amplifiers are described in the Transmitters chapter.

PSK

The popular HF real-time data mode PSK31 has been used successfully on the low frequency bands, as has PSK08 which runs at a quarter of the speed with consequent reduction in bandwidth and an increased signal/noise ratio. Both can be sent and received using *SpecLab* (see Appendix 2) and your computer's soundcard. PSK08 contacts have been made between the UK and Germany with copy reported as slightly better than normal speed CW.

Wolf

Wolf (Weak signal Operation on Low Frequency) was originally developed by KK7KA from techniques used for communication with deep space probes, and adapted especially for amateur LF operation.

Wolf works by encoding a 15 character alphanumeric message as 960 bits using convolutional coding techniques. The resulting massive redundancy in the transmitted signal allows sophisticated error-correcting techniques at the receiving end to reconstruct the original message even if a large proportion of the transmitted bits are corrupted by noise and interference. The signal is transmitted using phase shift keying at 10 bits per second, so the message takes 96 seconds to transmit.

Further enhancement of weak signal recovery is achieved by repeatedly transmitting the same message; a further 16 messages can be integrated together at the receiver to improve the signal to noise ratio at the expense of increasing the time required to transmit and receive a message up to about half an hour (this may seem a long time, but it compares well with the QRSS speeds typically used for intercontinental working).

The receiving software attempts to update and decode the signal every 96 seconds, producing the display in **Fig. 10.4**.

The original Wolf software [9, 10] utilised 'off-line' processing, where the transmitted signal was first encoded, then played back as an audio recording via

```
C:\wolf>wolf -q 10419d.wav -r 8018.527 -f 800.108 -b 5000 -c 7
WOLF version 0.61
t:   24 f:  0.592 a:  0.3 dp:115.6 ci: 9 cj: 85 9O/6R/1R3BK .DO ?
t:   48 f:  0.247 a: -0.9 dp:113.1 ci:12 cj:215 GGE7MXEPSIHAE28 -
t:   96 f:  0.590 a:  0.5 dp:109.2 ci: 7 cj: 49 7IN937BME53EEAQ ?
t:  192 f: -0.068 pm:   609 jm:602 q: -15.6 -7.6 T7.WE/7LIJWS4/0 ?
t:  288 f:  0.000 pm:   868 jm:796 q: -11.3 -7.6 Q3K*0X0U*FC195Q ?
t:  384 f:  0.000 pm:  1463 jm:796 q:  -9.4 -7.8 R6FFJWB27/2FR88 ?
t:  480 f:  0.000 pm:  1739 jm:796 q:  -9.5 -7.5 G1XLRUIT5MD26L2 ?
t:  576 f:  0.000 pm:  2228 jm:796 q:  -9.3 -8.2 RZMD5AGU4QOXDG  ?
t:  672 f:  0.000 pm:  2749 jm:796 q:  -7.6 -7.9 QV0SB/5BD.14XJ. ?
t:  768 f:  0.000 pm:  3319 jm:796 q:  -6.3 -8.6 88F4*KHBLL2AI83 ?
t:  864 f:  0.000 pm:  3957 jm:796 q:  -5.6 -6.6 5A7F7HOBLQ6OTOT ?
t:  960 f:  0.000 pm:  4682 jm:796 q:  -4.7 -5.3 M0BMU 10MW ERP  -
t: 1056 f:  0.000 pm:  5349 jm:796 q:  -3.6 -3.9 M0BMU 10MW ERP  -
t: 1152 f:  0.000 pm:  5747 jm:796 q:  -3.1 -3.1 M0BMU 10MW ERP  -
t: 1248 f:  0.000 pm:  6325 jm:796 q:  -2.4 -2.9 M0BMU 10MW ERP  -
t: 1344 f:  0.000 pm:  6944 jm:796 q:  -1.9 -2.4 M0BMU 10MW ERP  -
t: 1440 f:  0.000 pm:  7637 jm:796 q:  -1.5 -1.7 M0BMU 10MW ERP  -
t: 1536 f:  0.000 pm:  8520 jm:796 q:  -1.1 -1.0 M0BMU 10MW ERP  -
t: 1632 f:  0.000 pm:  8898 jm:796 q:  -0.3 -0.4 M0BMU 10MW ERP  -
```

Fig 10.4: WOLF uses a different approach to low frequency DX working. After several transmissions (one per line) are processed, the system has accumulated enough information to lock onto the signal and decode the message text, shown on the right

the PC sound card. The received signal was recorded, again using the PC, and the complete recording processed afterwards to extract the signal.

Tests using this software achieved the first trans-Atlantic reception of a digital mode on 136kHz. A more recent 'Windows GUI' version of the Wolf software, by DL4YHF and KK7KA [11], greatly simplifies operation by performing all signal generation, reception and processing in real time and using a standard Windows interface on the PC.

Successful Wolf operation requires very good frequency stability (a small fraction of a hertz) and careful setting up and calibration of frequency errors in the transmitter, receiver and PC sound card using the utilities built into the software. It is not a mode to be attempted by beginners.

A Wolf signal requires a bandwidth of nominally 10Hz. Interference from Loran sidebands, which are 8Hz apart, can be a problem in western Europe. The weak-signal capability is comparable with QRSS60, but Wolf has the advantage that the speed of transmission is several times greater and is therefore able to take advantage of shorter periods of good propagation conditions.

Because of the need for a linear amplifier and the setting-up complexity, Wolf is rarely used. However, it is arguably the best mode so far used for working intercontinental DX on the 136kHz band. WSPR-15 may come close but it is only available as a beacon, so two-way contacts are not possible.

Better synchronisation

Some of the work of decoding data signals is in gaining and maintaining synchronisation between the transmitting and receiving stations, ie working out when each code element starts and finishes and 'locking' the send and receive frequencies (even phases) together. Peter Martinez, G3PLX, experimented with using cheap Global Positioning System (GPS) modules to derive a 1 pulse per second to calibrate the soundcard, receiver and demodulator to a very accurate degree. One application of this, by ZL2AFP, is Clicklock2 [12].

Modes compared

Table 10.2 shows the modes most commonly used on the LF and MF bands. Your choice of which mode(s) to transmit will depend on the limitations of your transmitter and whether you want to chat or beacon. Since time and signal to noise ratio are linked, you will need to decide whether you want to work local stations rapidly, or intercontinental DX over several hours, or something in between.

Of course, you are not restricted to what your transmitter is capable of doing. All of these modes can be decoded using the station computer, and many allow 'listeners', whether licensed or not, to provide useful reports via an internet connection.

More modes

As many low frequency enthusiasts are compulsive experimenters, narrowband modes are constantly being developed and tried out. There has even been a two-way 'slow speech' contact using digitally slowed-down audio [13].

News of new, developing and experimental modes can be found on the RSGB_LF-Group [14].

CHAPTER 10: MODES USED ON 136kHz AND 472kHz

Mode	QSO mode available	Used in beacon mode	Linear transmitter needed	Multi-freq	Multi-pass	Time to send 'G3XDV' (s)	Approx bandwidth (Hz)	Web reporting / map	Tx/rx oscillator stability	Accurate timing required
CW (12WPM)	yes	rarely	no	no	no	7	10	no	low	no
QRSS3	yes	rarely	no	no	no	186	0.4	no	medium	no
QRSS10	yes	rarely	no	no	no	620	0.1	no	high	no
QRSS30	rarely	yes	no	no	no	1860	0.03	no	very high	no
QRSS60	no	yes	no	no	no	3720	0.017	no	very high	no
QRSS120	no	rarely	no	no	no	7440	0.008	no	very high	no
DFCW3	yes	rarely	no	yes	no	75	4.5	no	medium	no
DFCW10	yes	rarely	no	yes	no	230	1.5	no	high	no
DFCW30	rarely	yes	no	yes	no	690	0.6	no	very high	no
DFCW60	no	yes	no	yes	no	1380	0.3	no	very high	no
DFCW120	no	yes	no	yes	no	2760	0.15	no	very high	no
SMT-Hell	yes	rarely	no	yes	no	68	8	no	high	no
Jason	yes	no	no	yes	no	13	6	no	medium	no
PSK08	yes	no	yes	yes	no	7	10	no	medium	no
PSK31	yes	no	yes	yes	no	2	38	no	medium	no
Opera4	no	yes	no	no	no	240	2	yes	medium	no
Opera8	no	yes	no	no	no	480	0.6	yes	high	no
Opera32	no	yes	no	no	no	1920	0.2	yes	high	no
WSPR2	no	yes	no	yes	no	120	6	yes	medium	yes
WSPR15	no	yes	no	yes	no	900	0.8	yes	high	yes
WOLF	yes	yes	yes	yes	yes	Note	10	no	extremely high	yes

NOTES:
- QSO mode is of course 'available' when using dot lengths of 60 and 120 seconds, but these modes are useful only on long DX skywave paths which are not open long enough for a QSO at this speed.
- Opera4 is used only on the 472kHz band.
- Opera32 is used only on the 136kHz band.
- Wolf relies on repeated transmissions so the time to receive a callsign varies.

Table 10.2: LF modes compared. Use this chart to find the mode that suits your requirements and the restrictions of your station

Fig 10.5:
Connecting the receiver's audio output to the sound card input

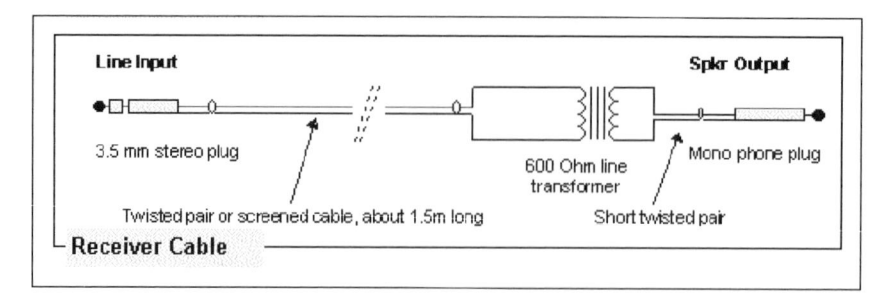

Interfacing with a computer

To operate virtually any LF/MF mode (the exception is normal speed Morse), a modern computer with a sound card is required. The computer specification depends on the mode and its software, but even a 233MHz Pentium will provide a lot of pleasure.

Some circuitry is also needed to interface the computer with the radio. Its complexity will depend on the whether simple on-off keying (or dual frequency FSK) or multifrequency audio is to be used.

The connections between the computer and the transceiver are quite straightforward, and most amateurs should be able to build the required cables.

A brief outline of interfacing is shown below. More detailed information can be found at [15].

Fig 10.6:
Connecting the sound card's output to the transmitter's audio input

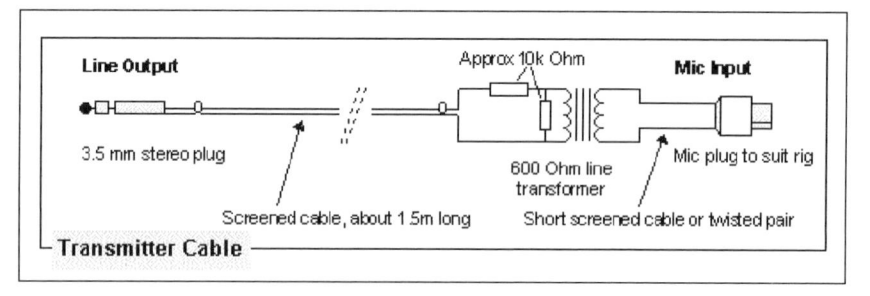

Receive audio

Whatever mode is in use, a connection must be made between the receiver's audio output and the computer's sound card input. **Fig 10.5** shows a suitable arrangement. The transformer is important (see below). If your radio has a socket that provides a fixed-level audio output, this would be preferable to using the headphone jack where the output volume must be kept constant. The sound card on a laptop may not have a line input so the microphone input will have to suffice.

Transmit audio

If you are using a audio tones to drive the transmitter, a similar interface needs to be constructed to connect the sound card output to the transmitter's audio input (**Fig 10.6**). The resistors in the transmit cable are used to attenuate the sound card signal so that it does not overload the transceiver. If the microphone socket is used, a lower value of resistor may be necessary across the trans-

CHAPTER 10: MODES USED ON 136kHz AND 472kHz

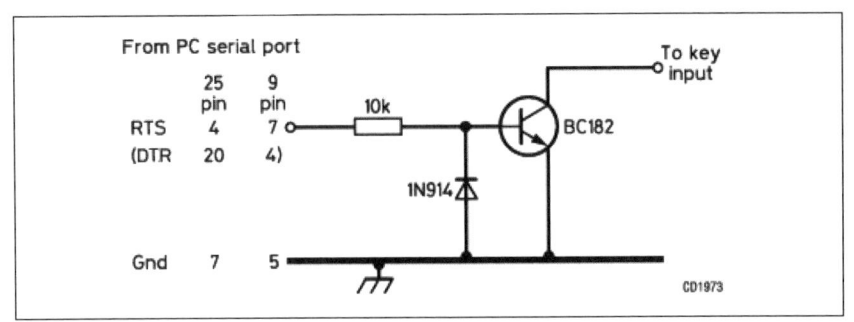

Fig 10.7: ON7YD's suggested simple interface for his *QRS* sending program. A capacitor could be added across the diode to decouple any RF

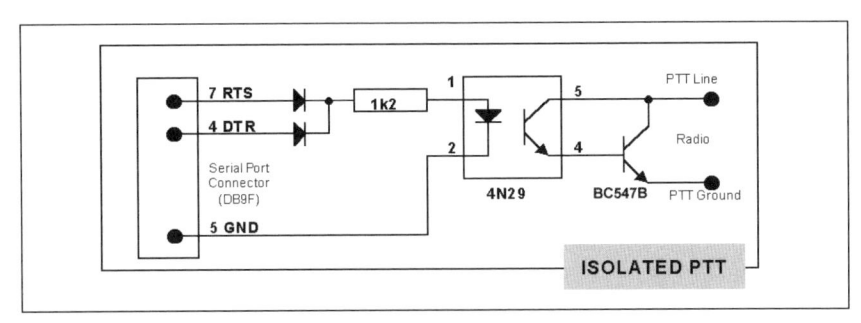

Fig 10.8: An improved keying interface with optical isolation

former to further attenuate the audio; if an accessory socket is used, the values shown should suffice.

The importance of isolation

The transformers shown in Figs 10.5 and 10.6 provide complete DC isolation between the computer and the radio transceiver. The most compelling reason to do this is to prevent serious damage to the radio and computer.

Most power supplies are grounded for safety reasons. If the power supply cable to the transmitter becomes loose, the full transmitter current can pass through the microphone circuit, down the cable and through the computer sound card to ground via the PC power cable. Even if the transmitter power cable is considered reliable, significant current could still flow through the sound card cable, causing instability, hum and RF feedback. The simple expedient of isolating the connections also reduces the risk of RF in the computer, and computer noises in the radio.

Keying and PTT circuits

Unless VOX/break-in is used, any mode will need a way to switch the transmitter on and off under control of the computer.

Modes that do not involve audio will also need a means of operating the transmitter's keying circuit and possibly a means of toggling between two frequencies (for DFCW).

Fortunately, each of these functions can be carried out by using a version of the same circuit, though the detail may have to be modified for your particular transmitter arrangement. **Fig 10.7** shows a simple interface by ON7YD. A better arrangement uses an opto-isolator to protect both computer and transmitter (**Fig 10.8**). This circuit will work for a transceiver with positive voltage on the PTT line and a current when PTT is closed of up to 100mA.

The digital mode software usually controls the transceiver via a serial port, by driving RTS or DTR (often both) positive on transmit, with an appropriate delay before sending tones out from the sound card (or keying in the case of Opera). Modern computers, particularly laptops, may not be equipped with a serial port. Interfaces are available cheaply which convert a USB port into a virtual serial port.

Soundcard controls

Recording and playback levels, as well as input / output selection are made using the computer's own operating system, and invariably accessed on Windows machines by clicking on the loudspeaker icon in the lower right taskbar. This usually brings up a 'Sound Mixer' or 'Volume Control' window. The menu options vary between manufacturers, but it will always be possible to select between all the input and output connections, to set recording and playback (volume) levels and to select which soundcard is the default. Hidden settings, or those not on immediate display, can usually be found in 'Advanced' or the Record menu options.

References

[1] ON7YD on QRSS: *http://www.qsl.net/on7yd/136narro.htm*
[2] PSKreporter: *http://pskreporter.info/pskmap.html*
[3] Opera mode: *http://groups.yahoo.com/group/O_P_E_R_A_/*
[4] WSPR software: *http://physics.princeton.edu/pulsar/K1JT/wspr.html*
[5] WSPRnet: *http://wsprnet.org/*
[6] SpecLab: *http://www.qsl.net/dl4yhf/spectra1.html*
[7] G3YXM on Jason: *http://www.wireless.org.uk/jason.htm*
[8] Jason software download: *http://www.weaksignals.com/*
[9] WOLF for beginners: *http://lowfer.us/k0lr/wolf/wolf4beginners.htm*
[10] WOLF software: *http://www.scgroup.com/ham/wolf.html*
[11] WOLF for Windows: *http://www.qsl.net/dl4yhf/wolf/*
[12] Clicklock2: *http://homepages.ihug.co.nz/~coombedn/FILES/Clicklock2.zip*
[13] Slow speech experiment: *http://www.qru.de/slowvoice.htm*
[14] RSGB LF Group: *http://uk.groups.yahoo.com/group/rsgb_lf_group/*
[15] 'Computers in the Shack', chapter by Andy Talbot, G4JNT, *Radio Communication Handbook, 11th ed,* RSGB

11

VLF - below 30kHz

In this chapter:

- What can be heard below 30kHz
- Receiving equipment
- Modes
- Receive antennas
- Permission to transmit
- Transmitters
- Transmit antennas

IN RECENT YEARS, AMATEURS attempting to push back the boundaries of experimentation have obtained special permits to transmit in the Very Low Frequency (VLF) range. It may come as surprise to learn that the area below 30kHz already has many commercial and military radio stations, and that this region has been used for worldwide communication using very high power transmitters and immense antennas.

Amateurs of course do not have access to high power and vast antenna farms, so is there any point in doing such experimentation? Surely, the signal will barely make it to the garden fence. Well, that's what the "experts" said about the old 73kHz band where the best one-way DX was eventually UK to Alaska!

Undaunted by such pessimism, stations in Germany, Holland, Austria, the Czech Republic and the UK have all transmitted on the band and some have been received several hundred kilometres away.

What can be heard below 30kHz

Fig 11.1 shows a spectrogram of part of this frequency spectrum with the callsigns marked. These are all high power stations and many are used for communicating with submarines as VLF can penetrate water to a depth of tens of metres. Several transmit narrowband data or carriers, but occasionally Morse code can still be heard.

Every few months, a special Morse code transmission takes place on 17.2kHz from the Swedish station, SAQ. A purely mechanical transmitter is used at the historic Grimeton site and this results in a

Fig 11.1: Using *SpecLab* to receive the spectrum from 12 - 26kHz. Strong commercial stations can be seen at the high frequency end

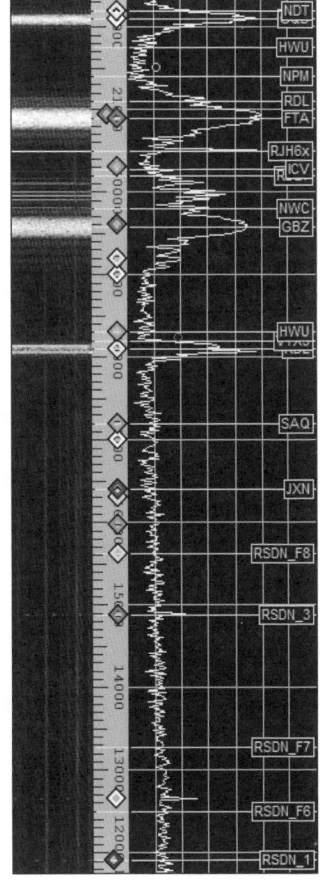

169

characteristic slightly unstable note. The signal is strong enough to be heard easily over much of Europe and by well equipped stations in the USA. Sending periods are announced on their web site [1], and will be mentioned on the RSGB_LF_Group [2].

Static can be a major problem at VLF, especially during the summer months. Harmonics of the mains electricity supply can be an issue, as can interference from domestic equipment.

A good test of your receive system is to listen on 11.905 kHz, 12.649 kHz and 14.881 kHz where the Russian Navy "Alpha" stations [3] are located. These appear to pulse on and off every second or so with the same frequency shared by several stations across Russia. If you can hear these at all, that is a good start, but if you can hear all of the pulses at good strength you have a good receiver. They can just be seen on the wideband plot of Fig 11.1. Note that the Alpha stations are frequently off the air, so some patience is required.

Most commercial activity takes place above 11kHz, whilst amateur transmissions have kept below 9kHz, sometimes as low as 5kHz, to keep within local licensing conditions.

Some amateurs run grabbers (see the chapter on Operating Practice) for VLF which can give a useful indication of activity, both commercial and amateur [4]. Amateur operation is very sporadic, so it is important to keep an eye on these grabbers and on any announcements on the RSGB_LF_Group [2].

Much more information on this fascinating part of the spectrum can be found in a book by IK1QFK [5] and on Paul Nicholson's web site at [6].

Receiving equipment

Most receivers do not extend below 30kHz. So how do you listen? One option is a selective measuring set (see the chapter on receivers), and some software defined radios (SDRs) will tune down almost to DC. However, it is important to ensure that they have the very high frequency stability required for amateur transmissions.

The most popular receiver is also the cheapest if you have a computer with a sound card. Simply connect the antenna to the audio input of the sound card, then download and run the free *Spectrum Laboratory* (*SpecLab*) software [7] which has several built-in options for VLF reception (see the Load Settings menu). A full description of how to set up *SpecLab* is beyond the scope this book, but help is always available from other amateurs at [2]. Accurate frequency setting and stability is not a problem as SpecLab can easily be locked to one of the strong commercial stations. It is important to ensure that your sound card or computer is protected from damage by lightning static. External sound cards can be obtained very cheaply in the form of USB dongles; these often work quite satisfactorily and can save your built-in card from accidental damage.

Modes

It has already been mentioned that low frequency amateur signals are very much weaker than commercial ones. This is even more the case at VLF. A naval station is likely to run a megawatt to a vast antenna array some 300m high, whilst the best amateur station has run a few hundred watts to a kite borne antenna.

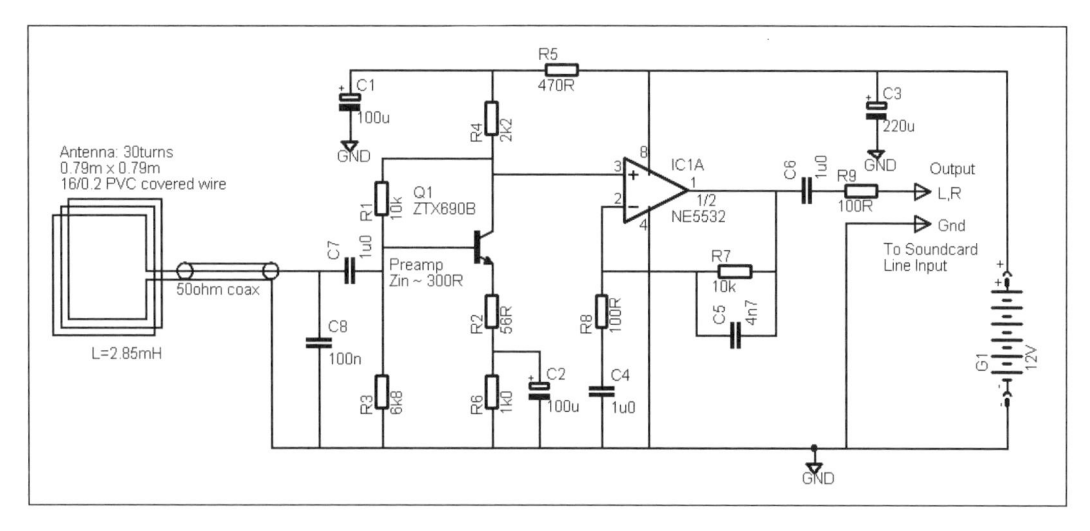

Fig 11.2: M0BMU's preamplifier works well with a loop antenna. He used this system to receive DK7FC on 8.97kHz. Q1 can be replaced by a BC337 with little change in performance

It is obvious, then, that something must be done to increase efficiency, and as with the 136kHz band the solution is to use QRSS (or DFCW). It is common to have dot lengths of an hour or even several hours, and a receive bandwidth of a few tens of microhertz. This dramatically improves the signal to noise ratio, but it makes it essential to have an extremely high degree of frequency setting and stability. It also restricts the information that can be transmitted to little more than a single Morse letter. A few stations located close to each other have used faster modes, but for most purposes VLF operation takes a lot of patience.

Receive antennas

Most Marconi-style wire antennas will provide enough signal to overcome the internal noise at VLF. It is quite possible to use an untuned wire (or a wire tuned for another frequency such as 136 or 472kHz) but for best performance it may help to add inductive loading - several millihenries - to tune the antenna to the correct frequency. For purely receive purposes the inductance need not be physically large, and could be tuned by adding a series capacitor.

A loop antenna can be used with the preamplifuer in **Fig 11.2**.

Permission to transmit

The legality of transmitting below 9kHz varies from country to country. Some regard this part of the spectrum as unregulated, others require special permission, a few will default to a refusal. In the UK, a Notice of Variation must be requested by writing to the regulator, Ofcom [8]. A few UK stations have successfully applied for such an NoV in recent years.

Transmitters

There are no amateur radio transmitters available for VLF. However. many high power hi-fi audio amplifiers can easily be found at relatively low prices, so it is easy to run several hundred watts output. Actual radiated power will, of course be very low - measured in microwatts.

For purely local experiments, a signal generator or basic audio oscillator can be used to drive the power amplifier. But 'DX' working requires a very high degree of stability so that a coherent carrier may be transmitted for hours at a time. Sometimes, DFCW may be used to make the amateur signal stand out from carriers produced by man-made interference.

The impedance of an amateur antenna is likely to be very high (perhaps a kilohm), whilst the output of an audio amplifier will be very low (a few ohms). Some matching is required. This often takes the form of a step-up transformer wound on ferrite salvaged from an old CRT television.

Transmit antennas

As with 136/472kHz the preferred antenna is the Marconi vertical, brought to resonance with an inductance. Most stations have used their already established LF antennas but DK7FC has carried out some successful tests with a kite-borne vertical. As with any LF antenna, the more wire in the air, as high as possible, is the rule.

Because even the largest amateur antenna will be extremely small compared to the wavelength, a very big inductance is needed to bring it to resonance.

Antenna current will be no more than a few hundred milliamps, but be prepared for extremely high voltages on the loading coil and the antenna (tens of kilovolts). This requires very good insulation and great attention to safety to avoid injury or fires.

DF6NM's 350mH loading coil wound on numerous buckets, insulated by Lego bricks. Tuning is by adjusting the overlap between the buckets

G3XBM has carried out ground loop experiments whereby he feeds RF current into a pair of electrodes driven into the ground a few hundred metres apart [9].

Summary

It is easy to monitor this interesting part of the spectrum at low cost. Receiving amateur transmissions requires a little more effort but those doing the transmitting will really appreciate your report.

Transmitting stations are few and far between, and transmitting sessions are infrequent, so it is essential to find out when to listen and on what frequency. The effort is worthwhile, though, as receiving a distant signal on such a low frequency can provide a real buzz.

References

[1] SAQ Grimeton: *http://www.grimeton.info/*
[2] RSGB LF Group: *http://uk.groups.yahoo.com/group/rsgb_lf_group/*
[3] Alpha stations: *http://en.wikipedia.org/wiki/Alpha_(radio_navigation)*
[4] VLF Grabbers: *https://sites.google.com/site/sub9khz/vlf-grabbers*
[5] *Radio Nature*, Renato Romero, RSGB
[6] Paul Nicholson: *http://abelian.org*
[7] Spectrum Laboratory: *http://www.qsl.net/dl4yhf/spectra1.html*
[8] Ofcom: *http://www.ofcom.org.uk/contact-us/*
[9] G3XBM: *https://sites.google.com/site/sub9khz/earthmode/vlf-xbm-blog*

Appendix 1

Information sources

In this appendix:

- Web sites
- RSGB LF Group
- Publications

EACH OF THE CHAPTERS in this book has a list of references, mostly web sites but also books. In addition, there are many resources available - most of them free - for those seeking more about low frequency communication. Those that give the best information at the time of publication are listed below, but new web sites and publications become available all of the time. Of all of these references, the most useful is the RSGB LF Group on the internet.

Web sites

Operators

The following sites are operated by individuals and organisations with practical experience of LF work. They are all well-established and it is hoped they will be accessible well into the life of this book. As with all information on the world wide web, their content is not edited and cannot be guaranteed to be one hundred per cent accurate. Some are also quite old, but much of the information is still current, or at least interesting. Most are well worth a look and will provide the curious experimenter with a host of solutions and many more ideas.

In addition to the topics listed below, most individuals include a description of their station and a log of their best contacts.

DF6NM (*http://www.df6nm.de/*): VLF, LF and MF band monitoring using a directional spectrogram, and Loran-C reception charts.
DK8KW (*http://www.qru.de/*): QRSS intro, 'slow voice' experiments, VLF, earth electrode propagation, selective level meters, DDS, LF utility stations, dBm to dBµ conversion chart, calibrating receiving equipment.
DL4YHF (*http://www.qsl.net/dl4yhf/*): Masses of material from the author of *SpecLab*. Software, PIC projects and operating info.
G0MRF (*http://www.g0mrf.com/*): Many projects for LF and MF, and expedition reports.

G3NYK (*http://www.xeropage.co.uk/g3nyk/*): A mine of experimental data and theories on LF propagation. Also a history of Rugby Radio, and lots on time and frequency.

G3XBM (*https://sites.google.com/site/g3xbmqrp3/*): Antennas and transverters for LF and MF, details of VLF experiments, and earth-antennas.

G3YMC (*http://www.davesergeant.com/*): Descriptions of practical experiments on LF loop and Marconi antennas in a very small garden, MF experiments and television receivers as generators of QRM.

G3YXM (*http://www.wireless.org.uk/index.htm*): After the *rsgb_lf_group* (see below), this is the place to go for news. Also featured are many projects including transmitters and pre-amplifiers, and advice for beginners on both LF and MF. Articles include expedition reports. The Matrix is a database of who made the first 136kHz between any two countries.

G4JNT (*http://www.g4jnt.com/*): A mass of technical information and projects, including synthesizers, GPS locking and LF/MF transmitters,

G4WGT (*http://myweb.tiscali.co.uk/wgtaylor/index.html*): Construction and operating information for the 136kHz and 472kHz bands, grabbers and VLF experiments.

GW3UEP (*http://www.gw3uep.ukfsn.org/*): Lots of simple projects for 472kHz, including transmitters and test gear.

IK2PII (*http://www.qsl.net/ik2pii/*): The many 136kHz projects include the design and construction of a 200W transmitter, a direct conversion receiver and a dedicated QRSS receiver.

OK1FIG (*http://ok1fig.nagano.cz/136k.htm*): Pictures of equipment for 136kHz, and expeditions. Recordings of many LF stations in mp3 format. Many screen shots of LF stations using QRSS, DFCW and Hell. Transmitter projects and much more.

ON7YD (*http://www.qsl.net/on7yd/136khz.htm*): Lots of information on LF antennas and low bandwidth modes. Many links to articles on LF as well as recommended literature.

RU6LA (*http://136.73.ru/*): Mostly Russian language, but some English including a history of LF radio and list of LF DXpeditions.

S52AB (*http://lea.hamradio.si/~s52ab/*): LF, MF and NDBs

SV8CS (*http://sv8cs.blogspot.gr/*): LF and MF activity from Greece.

VE7SL (*http://members.shaw.ca/ve7sl/*): Canadians on 136kHz, advice for beginners and transpacific tests by VA7LF..

VK1SV (*http://people.physics.anu.edu.au/~dxt103/136/*): 136kHz from Australia. Antennas, variometer, transmitter, earth antenna experiments and a grabber.

VO1NA (*http://www.ucs.mun.ca/~jcraig/lfex.html*): Operational details from the owner of the strongest signal from the Americas into Europe.

W3EEE (*http://www.hifidelity.com/w3eee/*): Evaluating receive antennas. Transatlantic LF DX activity from the US end.

W4DEX (*http://www.w4dex.com/*): Details of W4DEX's experimental LF and MF activities as WD2XKO and WD2XSH/10.

WD2XSH (*http://www.500kc.com*): The 'official' ARRL Six Hundred Metre Experimental Group site detailing activity in the US 505 - 510 kHz experimental allocation.

APPENDIX 1: INFORMATION SOURCES

Other sites

http://www.wireless.org.uk/g4fgq/index.html
Archive copy of the former web site of the late Reg Edwards, G4FGQ. Computer programs, mostly operating under DOS, including the useful TOROID.EXE.

http://www.amqrp.org/projects/Toroid Design Tool/toroiddesign.htm
Page where you can enter a toroid type and the required inductance and the number of turns will be calculated.

http://www.dxlc.com/solar/
Solar information, useful for propagation prediction, from the DX Listeners Club. In German and English.

http://www.hfradio.org/propagation.html
Comprehensive solar and ionospheric data, updated daily.

http://sec.noaa.gov/
Much propagation data from the National Oceanic and Atmospheric Administration (NOAA)

http://lasp.colorado.edu/space_weather/dsttemerin/dsttemerin.html
Colorado University predictions of Dst data over the next few hours as well as archival data..

http://swdcwww.kugi.kyoto-u.ac.jp/dst_realtime/presentmonth/index.html
Kyoto University's monthly graphs of Dst data.

http://www.blitzortung.org/Webpages/index.php?lang=en&page=1
http://www.isleofwightweather.co.uk/live_storm_data.htm
Real-time map displays of lightning, both worldwide and UK. Useful to check the origin of QRN, or simply abandon that sked!

http://www.amrad.org/projects/lf/
The LF pages of AMRAD, a group of keen US experimenters. Includes various low frequency projects and info.

http://www.lwca.org/
The web site of the Long Wave Club of America; the low frequencies from a US perspective, including a message board, news and solar activity.

RSGB LF group

Most of those who are active on the low bands worldwide are 'members' of the RSGB LF Group. Since the mid-1990s this forum has facilitated the coordination of the very latest activity and and the exchange of technical discussion to all who subscribe (free). Originally hosted by blacksheep.org, it became a Yahoo group in June 2012.

Membership is free and open to all. There are two ways to join:

- Send an e-mail to: rsgb_lf_group-subscribe@yahoogroups.co.uk, or
- Go to the web site: *http://uk.groups.yahoo.com/group/rsgb_lf_group/* and follow the appropriate link.

You can choose whether to receive each message as an e-mail, or as a daily digest of e-mails, or view them on the web. Posting a message can be done as an e-mail or on the web site.

There is no junk mail, just news and comment, plus technical discussion and advice. This is the place for technical discussion, to learn about forthcoming activity and new narrowband modes, or to ask for skeds. It is also where you can ask questions on LF matters, no matter how simple or complex, with the assumption that you will get replies from experienced, knowledgeable and friendly people.

As with all Yahoo forums, there is a searchable message archive, photos, files and useful links.

Publications

Although it was first published in the first half of the last century, *The Handbook of Wireless Telegraphy 1938 Vol II*, is still the place to find useful theoretical and practical information. Published by the Admiralty and reprinted each year for a decade, this tells how LF was used on board ships. Strangely, because of the limited space available on a warship the problems were similar to those facing radio amateurs. Copies are still available from time to time in secondhand shops and on internet auction sites.

There are very few books about amateur radio on the low frequency bands. However, books that are not solely about this part of the spectrum can also contain much useful information.

Out of print publications that are still likely to be available secondhand include: *Digital Modes for All Occasions* by Murray Greeman, ZL1BPU, which includes much simply explained theory and practice for those interested in using some of the modes described in this book; and *The Antenna Experimenters Guide* by LF enthusiast Peter Dodd, G3LDO deals with HF antennas but has much to offer anyone keen to understand more about how antennas work.

The RSGB publishes: *Elimination of Electrical Noise* (RSGB) for information on tracking down noise sources and dealing with your neighbours, and the 800-page *Radio Communication Handbook*, which is not only essential for a general grounding in practical amateur radio but also includes a chapter on the frequencies below 1MHz. These are available from the RSGB, 3 Abbey Court, Fraser Road, Priory Business Park, Bedford MK44 3WH; tel 01234 832700; http://www.rsgbshop.org/.

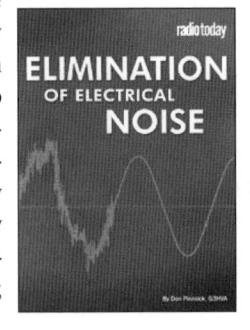

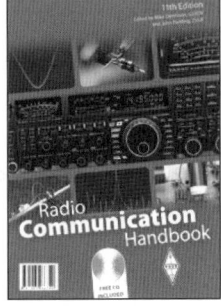

Lastly, the RSGB's monthly magazine, *RadCom*, has an LF column in every other edition, covering activity on VLF, 136kHz and 472kHz.

Appendix 2

Components and software

In this appendix:

- Capacitors
- Litz wire - is it useful?
- Ferrite and iron-dust cores
- Component sources
- Software

TRANSMITTERS FOR THE LOW frequencies may well run several hundred watts, particularly on the 136kHz band. Antennas can have very large voltages and/or currents, greatly in excess of those found on higher bands. These make it important to use high quality, properly rated, components to avoid overheating, fires or even explosions. At the same time, LF/MF antenna efficiencies are very low so losses must be minimised by using the correct materials.

The improved antenna efficiencies available at 472kHz makes it less important to use highly rated components, though it is still worth engineering your station properly.

Capacitors

Use capacitors with a high voltage and current rating for most RF power applications. Do not use components straight from the junk box as most types are likely to heat up, especially disk ceramics. For LF/MF, the following types of capacitors are best for high Q:

Ceramic: CG0 or NP0 types, usually restricted to less than 1nF and rated 50V or 100V. Other types of dielectric have high RF losses. Transmitting types are also available with much higher ratings, but are much rarer and expensive.

Mica: Silvered mica up to 10nF or so, rated at 350V or 500V DC are quite widely available and moderately expensive. Occasionally, surplus transmitting mica types are found with much higher ratings; these can deteriorate if very old.

Polystyrene: Up to 10nF or so, usually with voltage ratings of 160V or less are widely available and fairly cheap. Occasionally found with ratings of 1kV or more and values up to 1µF.

Polypropylene, Metallised Polypropylene: widely available in values from a few hundred pF to several µF, rated up to a few kV. Low loss at 136kHz and 472kHz. Best value for money, especially when large capacitance is needed.

Polypropylene types are the best choice in many LF/MF TX applications. If using surplus or second-hand components, be aware that polyester types look just the same, so check the part number on the manufacturer's web site, or test for RF loss, since polyester types have much higher loss. In RF applications, capacitor ratings are usually limited by internal heating due to the RF current, rather than voltage, so generous de-rating is often needed. Polypropylene and polystyrene types in particular should run no more than very slightly warm, because they are limited to quite low maximum operating temperatures.

Litz wire - is it useful?

When a radio frequency signal is passed through a wire, it travels on the outside of the wire. This is known as the skin effect and is the reason that VHF antennas can be made of tubing. It makes the resistance to RF greater than the DC resistance and and causes increased loss in inductors and loop antennas.

The solution is to use Litz wire, which is many *insulated* strands of thin wire bundled together to maximise the surface area. Second-hand Litz wire is sometimes available at rallies or from members of the LF community, although it is prohibitively expensive new.

Using Litz will certainly reduce losses - it can increase the Q of an air cored loading coil by a factor of two or three. Although the loss in the coil itself is reduced substantially, the effect on antenna efficiency may well be insignificant due to larger losses in the earth and the environment. However, reducing the loss in the coil also allows higher power operation before the coil overheats. For those who already have a highly efficient antenna system, it may be the next thing to try.

The older type of Litz has dark brown enamel insulation which is extremely hard to remove. Newer types have pink, red or orange 'self fluxing' insulation which can be removed by tinning in a solder pot, or with a hefty soldering iron. It is important to ensure that all strands of the wire are properly soldered; failure to do this will greatly increase losses and defeat the object of using Litz, as well as potential overheating.

Ferrite and iron-dust cores for LF and MF

Ferrite and iron-dust magnetic cores are familiar components in amateur HF construction projects, in the form of toroids (rings) and tuning slugs. They are also important components in most LF and MF projects. At low frequencies, as well as toroids, other forms of magnetic core are often used, such as transformer 'E' cores and pot cores. The material grades most suited to LF/MF construction are on the whole different to those used at HF. This section offers some general guidance about using ferrite and iron-dust cores at LF/MF; much more detailed information is available in the application notes at [1] and other manufacturers' web sites. Magnetic materials can be classified according to relative permeability (μ). This is basically the factor by which the inductance of a coil wound on a toroid core increases, compared to the same coil wound on a non-magnetic core. Iron dust materials have μ typically 10 - 100, while ferrites have much higher relative permeability, between about 100 and several thousand.

Iron dust cores are used in two general areas by LF/MF amateurs; low loss, high Q inductors for resonant circuits, and RF chokes for feeding DC current to PA stages, supply noise filters, etc. High Q inductors are generally required for transmitter tank circuits and output filters, for which the best material is the Micrometals -2 grade (usually colour-coded red). Toroids in this material are also widely used in the lower HF range, and are available from amateur radio component suppliers. A Q greater than 100 is usually achieved. Large toroid cores are used for LF, for example the T130-2 core is suitable for a 136kHz output filter up to a few hundred watts, and T200-2 up to 1kW or so. Smaller sizes such as the T68-2 are satisfactory for lower power levels at 472kHz.

RF choke applications can use 'power' grades of iron dust core. These cores have higher permeability than the -2 grade, and so achieve higher inductance with fewer turns of wire. RF losses are much higher though, so Q is usually less than 20, but this is usually an advantage for a choke because unwanted resonances are damped. The Micrometals cores of this type usually have a two-colour code, eg -26 grade is yellow/white, -52 grade is green/blue. This type of core can often be salvaged from scrap switch mode power supplies.

The main use for ferrite cores at LF/MF is as transformer cores. Different ferrite grades are best suited to different frequency ranges, with higher permeability materials being best suited to lower frequencies. For use in 136kHz PA output transformers and antenna matching transformers, ferrites with a permeability of around 2000 are best; these are the same grades that are used for switch-mode power supply transformers, and include Ferroxcube/Philips 3C8, 3C85, 3C90, 3C95 etc, Siemens/Epcos N27, N67, N87, Fair-Rite #75 and #77, Neosid F44 and many other similar materials. These are available as toroids, but more widely as transformer assemblies with two 'E' core halves and a plastic bobbin, and also pot cores. The higher numbers tend to be more recent developments with slightly improved characteristics, but differences between grades are not great. For 472kHz, the same cores can be used; the losses in the cores are higher, but since the power levels are normally lower, this is usually acceptable. The Fair-Rite #43 material is very satisfactory for 472kHz with somewhat lower permeability ($\mu = 850$), and lower losses. For small signal transformers, very high permeability cores can be used to achieve wide bandwidth with small windings. Low permeability ferrite cores (eg Ferroxcube/Philips 4C65, Fair-Rite #61) can be used to produce high Q small signal inductors, but these are not suitable for power applications due to non-linearity at high signal levels. Ferrite pot cores, provided with an air gap and a tuning slug for adjustment, are well suited to producing high-Q LF/MF inductors, but are not now widely available.

It is important to be able to either calculate the inductance, or the number of turns required for toroidal or other cored inductors and transformers as follows:

$L = N^2 A_L$, or $N = \sqrt{(L/A_L)}$

where A_L is the 'specific inductance' for the particular core, and N is the number of turns. A_L is usually specified by the manufacturers in nanohenries per turn, so the inductance value will also be in nanohenries.

If the A_L of a ferrite core is not known, it can be roughly estimated from:

$A_L = 4\pi \times 10^2 \times \mu A/l$ (nH/turn)

where μ is the permeability (a rough guess will often do for transformer design; most 'switch-mode' materials have μ in the range 2000 - 2500), A is the cross-sectional area of the core (in square metres) and l is the length of the

**Fig A2.1: Cores for LF/MF (l to r): Back row: EC41, EE42, ETD59 transformer core assemblies; PA output transformer for modified G0MRF 400W TX wound on LOPT core.
Front row: Three pot cores, T200-2 and T68-2 iron dust toroids, three assorted 3C85 and 3C90 toroids**

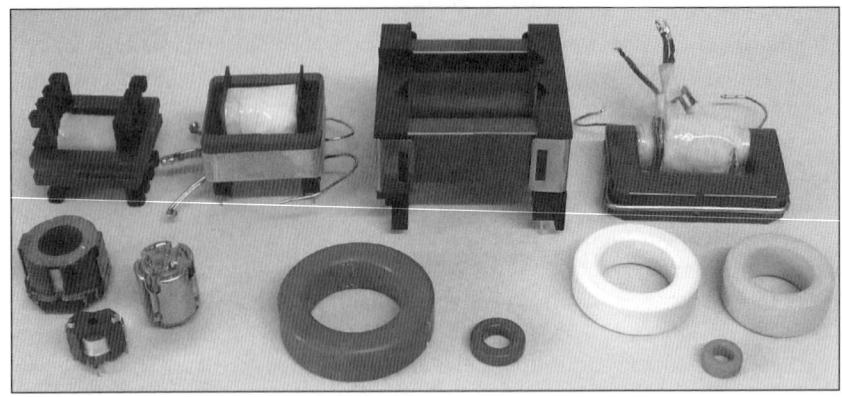

magnetic path through the core in metres (the average of the inner and outer circumference of a toroid). The value of A_L will typically be in the range of thousands of nanohenries per turn for cores likely to be used. Typically, the inductance of a transformer winding is chosen so the reactance will be 5 - 10 times the circuit impedance level at the lowest operating frequency.

A free program by DL5SWB [2] will calculate the characteristics of an unknown core from the measured inductance of a few turns of wire.

For power transformers, it is also important that the magnetic flux density in the core is not too high; a high flux density will lead to high losses, severe heating, and problems with core saturation. Peak flux density (measured in Tesla, T) of about 100mT - 150mT is usually a safe figure for these ferrites. For a sine wave RF voltage applied across a winding, the peak flux density, B_{peak}, can be calculated from:

$$B_{peak} = V_{RMS} / 4.44fNAe$$

Where f is the frequency in Hertz, N is the number of turns on the winding, and Ae is the effective cross-sectional area of the coil (in square metres).

More details on transformer design can be found in the chapter on transmitters, and at ferrite manufacturers' web sites [1].

'Switch-mode' ferrite cores are available from industrial component distributors such as RS components and Farnell (see below), who will also supply private customers. Transformer core and bobbin assemblies are available in several styles (eg EC, EE, ETD) and sizes (for example, an ETD 39 core assembly has overall length of 39mm). Small toroidal ferrite cores are also available from these sources, as well as from amateur radio suppliers; these are sometimes colour coded according to ferrite grade (3C85 toroids are usually a pink colour), but the modern tendency is to have either a plain white coating, or no coating at all.

The ferrite core from a line output transformer (LOPT) from a TV set or monitor can also be a useful transformer core. They are made from a 'switch mode' ferrite grade. The core can be recovered by carefully sawing away the original transformer windings; take care to remove any plastic spacers or adhesive separating the mating surfaces of the two core halves.

For receiver input filters, fixed inductors of the type manufactured as filter chokes can be used. For example the Panasonic 'ELC series sold by RS Components [3]. Note that they are not screened.

APPENDIX 2: COMPONENTS AND SOFTWARE

Component sources

Note that this information is believed to be correct at the time of publication. Some suppliers insist on a minimum quantity order so it may be useful to get together with others when ordering.

David Bowman, G0MRF (3C90, T200 and T130 toroids, DDS sources and LF/MF transmitter parts). The list of components and kits available is undergoing a re-vamp, but David will continue to support the LF/MF market. E-mail: *g0mrf@aol.com*. Web: *http://www.g0mrf.com/*.

Farnell Electronic Components Ltd (semiconductors, ferrites, chokes, capacitors) Comprehensive stock for commercial and amateur customers. Web: *http://www.farnell.com/*.

JAB Electronic Components (toroids, Toko coils, crystals, boxes, etc) Web: *http://www.jabdog.com/*.

Keytronics (semiconductors, crystals) 88 Hadham Road, Bishops Stortford, Herts, CM23 2QT.
E-mail: *pk@keytronics-uk.co.uk*. Web: *www.keytronics-uk.co.uk*.

Kitsandparts dot com (inductors, wire, toroids, filter kits) US-based company. Web: *http://www.kitsandparts.com*

Maplin Electronics Ltd (ferrites, components, boxes, connectors, wire) National Distribution Centre, Valley Road, Wombwell, Barnsley, South Yorkshire. S73 0BS. Tel: 0870 429 6000. High street presence in many towns countrywide. Huge printed catalogue available. Web: *www.maplin.co.uk/*.

Quartslab (crystals from 1.5MHz up)
Web: *http://quartslab.com/*

RS Components: (huge range of semiconductors, ferrites, capacitors, etc) UK Orderline: 08457 201201. Web: *http://uk.rs-online.com/web/*.

TLC Electrical Supplies (screw-ended earthing rods and couplers) Tel: 01293 565630. E-mail: *sales@tlc-direct.co.uk*. Web: *www.tlc-direct.co.uk*

Software

It is possible to operate on the low frequencies without using a computer, and several successful operators have done so. However, to make the fullest use of this part of the spectrum it is very useful to have a computer in the shack. For instance, the RSGB LF Group forum is the hub of the low frequency amateur radio community. The shack computer can also be used for making calculations and for using the various popular low bandwidth communications modes (see the Modes chapter). It is increasingly being used to control a software defined radio. Some recommended programs are listed below - all can be downloaded free from the Internet.

Calculators

These two programs are written to run under DOS, but they work fine in a 'DOS' window on a computer running Windows XP (and probably later versions). They are a bit fussy to use, but are worth a little trial and error.

TANT136.EXE (G4FGQ) (*http://www.wireless.org.uk/g4fgq/page3.html*): This is designed for those who want to use an 160/80/40m dipole with the feeders strapped to make an LF Marconi, although it can be used for purpose designed inverted-L or Tee antennas. Having input the size of the antenna, the feeder or vertical wire and the dimensions of the chosen coil former, the program calculates the coil winding details, the expected radiation resistance and the output power. TANT136.EXE should only be regarded as a rough guide, but it is a good starting point when building your first LF antenna.

SOLNOID3.EXE (G4FGQ) (*http://www.wireless.org.uk/g4fgq/page3.html*): Good program to use when designing air-core coils such as loading coils.

TOROID.EXE (G4FGQ) (*http://www.wireless.org.uk/g4fgq/page3.html*): This one calculates the number of turns required to make a specified inductance on a specified toroid. Very useful.

QRSS programs

With the exception of the keying software, all of these display an audio spectrum and use Fast Fourier Transform (FFT) to integrate the received audio. It is essential to have at least one of these if you want to receive QRSS and its variants. All require a sound card to be installed in the computer.

Argo by I2PHD / IK2CZL (*http://digilander.libero.it/i2phd/argo/index.html*): This has presets for most of the commonly used variations of QRSS, and very little is adjustable. This makes Argo the easiest of the FFT programs to use. Time and date can be displayed on screen. It will work with real time audio or a recorded (WAV) file, and will also save audio as a WAV file. Capturing 'snapshots' of screens, for a 'grabber' or simply for later analysis, is easy and may be automated on a timer (great for monitoring DX when you are in bed). Ideal for the beginner.

SpecLab by DL4YHF (*http://www.qsl.net/dl4yhf/spectra1.html*): The big daddy of spectrograms, Spectrum Laboratory is flexible and capable of many things that others cannot do. It is correspondingly complex to use. For QRSS use, the FFT, audio and display settings are fully configurable. There are many data export functions, including the ability to plot the strength of a signal and/or noise band. An audio spectrum analyser is included. For those with two antennas, the unique radio direction finder can be useful. The program will also cope with PSK08, Hell and other modes. For those whose transmitters have SSB drivers, SpecLab will output tones for QRSS, DFCW and various datamodes.

QRS by ON7YD (*http://www.qsl.net/on7yd/136narro.htm*): The most popular way to key a QRSS or DFCW transmission, this program covers a wide range of dot lengths from one second to one hour! From text input it will key the transmitter, and operate the PTT line if required, via the serial or parallel ports. A simple interface must be constructed for each line (see the chapter on operating practice). The QSK function allows you to listen between dots or dashes and an audible alarm sounds when the transmission is about to end.

Other narrowband modes

See the operating chapter for details of where to find the software for SMT Hell, Jason, Opera, PSK, Wolf, WSPR etc.

References

[1] Ferroxcube: *http://www.ferroxcube.com*. Fair-Rite Products Corp: *http://www.fair-rite.com*. Micrometals Inc: *http://www.micrometals.com*

[2] RS components: *http://www.rswww.com*

[3] *http://www.dl5swb.de/mini_ringcore_calculator.html*

Index

Activity levels / periods ... 141, 143, 150, 155
Antenna
    bandwidth 35, 45, 53, 56, 57, 62, 63
    current (see also Marconi antenna,
        and Loop antenna)......... 4, 23, 24,
        .... 42, 46, 47, 117, 124
    directivity....... 23, 52, 54, 56, 119, 130
    effective height (see Marconi antenna)
    efficiency......... 2, 24, 42, 45, 77, 121
    factor........................... 123
    feed impedance 5
    field strength 122
    loss resistance 24, 30, 119, 120
    measurement 116
    noise reducing 148
    positioning.................. 52, 53, 61
    radiation resistance.... 23, 24, 42, 45, 119
    120
    supports 48-50
    tuning (see also Inductors) ... 5, 9, 15, 33,
    .... 35, 36, 41, 43, 124, 126
Antennas........................ 23-66
    active whip.............. 52, 53, 61-62
    earth electrode 46-48, 172
    ferrite rod................... 62, 146
    for small gardens................. 31
    half-wave dipole 3, 119
    isotropic...................... 119
    loaded horizontal dipole 24
    Marconi (see Marconi antenna)
    loop (see Loop).................... 3
    matching 4, 37-39
    modifying HF antennas for LF/MF .. 3, 51
    quarter-wave ground-plane 3
    receiving (see also Loops) ... 5, 9, 14, 16,
    .... 33, 51-66, 171
    simple 3
    VLF (see VLF)

Bandplans (see Frequencies)

Beacons
    formal 143
    informal................ 143, 152, 160
    NDBs (see Non-amateur stations)

Commercial stations
    (see Non-amateur stations)
Components
    capacitors 43, 177
    ferrite and iron dust cores....... 178-180
    Litz wire 34, 179
    sources 181
Contacts, content of... 141-142, 155, 157, 158
Converter
    receive 15, 18, 19, 71-72, 78
    transmit (see Transverter)
Corona discharge............. 27-28, 32, 42
Countries
    on 136kHz.................. 7, 151
    on 472kHz................ 2, 7, 151
    on VLF 169

Distances achieved..................... 2
Dummy load (see Test equipment)....... 116

Earthing (see Ground system)
ERP (see Power)
EIRP (see Power)

Filter
    in receivers (see Receiver, bandwidth)
    low pass.............. 11, 75, 99, 106
Frequencies
    band limits................... 1, 2, 141
    bandplans 7, 143

GPS
    frequency locking 10, 76
    synchronising timing.............. 164
Grabbers (see Web reporting)

Ground system 24, 28, 29-30
    current in. 30
    earth stakes 5, 29
    effect on noise. 148
    radial wires 29, 30
    soil conductivity 29, 46
    water pipe 5, 29
    whip antennas 53

History. 1, 2

Inductors
    elevated, light weight 36
    ferrite tuned. 35
    Litz wire (see Components)
    loading. 5, 28, 31, 33-36, 37,
    . . . . 39-42, 172
    losses 34, 40-41
    measuring (see Test equipment)
    practical. 39
    tapped 35, 38
    variometer 4, 34, 35, 39-41
Information sources 173-176
    books. 176
    RSGB LF Group . 7, 8, 141, 149, 170, 175
    web sites (see also the References
        at the end of each chapter) 173
Interference
    cancelling 64
    from and to non-amateur stations
        (see Non-amateur stations)
    identifying sources 147-148
    man-made . . . 9, 51, 52, 53, 54, 63-66, 146
    170
    natural (static) . . 10, 6, 145, 157, 158, 170
    picked up on feeder. 55
    to other amateurs 42, 141, 143, 162
Inverted-L (see Marconi antenna)

Keying and PTT interface. 166-168

Licensing 1, 145
    permit to use 472kHz 7, 143
    permits to transmit on VLF 169, 171
    power limits (see Power)
Litz wire (see Inductors)
Loop antenna
    area 42, 44, 45, 56, 58
    bandpass 56, 58
    calculations 121, 122
    calibrated. 123
    directivity 24, 46, 146
    feedpoint. 43
    EWE 54
    for transmitting. 3, 24, 42-46

for receiving. . 5, 15, 16, 17, 46, 52, 53-54,
. . . . 55-61
    K9AY 54
    low-pass 56
    matching 44, 55
    practical. 44, 55
    shape 43, 45, 58
    thoughts on 56
    tuning 43, 53, 57, 60
    untuned 54
    voltages. 43
    Wellbrook 54
Loss
    environmental (non-earth). . 24, 27, 30, 31,
    . . . . 33, 122
    in loading coil (see Inductors)
    resistance in antenna system (see Antenna)
Luxembourg effect. 15

Marconi antenna 3, 23, 24-42
    capacitive loading wires 4, 24, 25,
    . . . . 31, 32, 120
    design considerations 25
    current distribution 3, 25
    height, effective, importance of 5, 25,
    . . . . 27, 29, 31, 120
    loading coils (see Inductors)
    matching 4, 9
    shape of. 5, 26, 31, 32
    voltage distribution 25
Measurement (see Test equipment)
Modes. 155-168
    bandplanning for (see Frequencies)
    categories of 155
    choosing the right. 152, 164-165
    CW 2, 7, 10, 15, 67, 69, 77, 78, 113,
    . . . . 141, 143, 150, 151, 155, 165, 169
    DFCW 71, 78, 143, 152, 153,
    . . . . 159-160, 165, 171, 172
    for VLF (see VLF)
    Hellschreiber 113, 162
    Jason 162, 165
    JT65 113
    narrowband 1, 6, 10, 162
    new 155, 164
    Opera 3, 7, 10, 15, 51, 69, 143, 149,
    . . . . 151, 152, 157, 160, 165
    PSK 150, 107, 113, 163, 165
    requiring linear amplification. . 71, 78, 107
    162
    ROS 143
    SSB (phone) 9
    QRSS . . . 3, 7, 10, 15, 67, 69, 77, 78, 113,
    . . . . 138, 142, 143, 151, 152, 153,
    . . . . 155-159, 161, 163, 165, 171

RTTY 71, 113	dead zone 130
SMT Hell. 162, 165	Disturbance Storm Time (Dst) . . 133, 134,
speed/bandwidth/sig-noise . . 151-152 156,	. . . . 135, 136
. . . . 158, 159, 160	experiments 160, 161
WOLF 107, 152, 163, 165	fading . . 129, 130, 136-137, 138, 139, 142
WSJT-X. 143	geomagnetic storm. . . . 133, 135, 136, 137
WSPR . . . 2, 7, 10, 15, 47, 51, 67, 75, 113,	good conditions, can we predict? . 135-136
. . . . 143, 150, 151, 152, 161, 162,	ground waves. 129, 137, 139, 150
. . . . 164, 165	in summer 138
Morse (see Modes, CW)	ionosphere 130-132, 137, 139
	long term changes 137
Noise, electrical (see Interference)	magnetosphere. 132
Non-amateur stations	on 136kHz. . . 2-3, 129-138, 150, 151, 152
as propagation indicators . . . 131, 136, 137	on 472kHz 2, 139, 142, 143, 150
. . . . 138, 139, 152, 170	Ring Current 133, 135
as tuning aids 6-7, 10, 14, 41, 157	solar cycle 137
frequency locking to (see also GPS) . . 170	solar flare. 132, 134, 135, 151
interference from . . 6-7, 10, 11, 15, 51, 64,	Sun, effect of 132-135
. . . . 145, 146, 147, 155, 162, 164	sky waves. 130, 136, 137, 142, 150
interference to 8, 143	Van Allen belts 133
navigation systems at LF. 129	X-ray flux 132, 135
NDBs 7, 139, 143	
on VLF 169	QRP operating 1
Opera (see Modes)	Receiver 9-22
Operating 141-154	AGC 10, 15
away from home 153	bandwidth / filters 10, 15
DX working 150	dynamic range. 10
on 136kHz 144-145	frequency calibration. 127
on 472kHz 142-143	frequency readout / steps. 10
Oscillator	frequency stability (see also GPS). 10
ceramic resonator 69	improving 14
crystal. 67, 68-69, 72, 100	intermodulation 14
direct digital synthesis (DDS). . 75-76, 162	non-amateur 12
frequency division 68, 69, 70, 71, 100	performance at LF/MF 6, 10
mixing. 69	preamplifier / preselector . . . 6, 10, 11, 14,
requirements 67	. . . . 15-18, 56, 57, 58, 62, 63, 78
signal-frequency 67	recommended / suitable. 6, 11
ultimate2 QRSS kit 76	remote 148, 149
variable (VFO) 67, 69-70, 89, 91	software defined (SDR) 6, 11, 13,
	. . . . 20-22, 170
Planning permission. 31	transceiver as. 6, 9, 11
Portable operating 58	vintage. 14
Power amplifier (see Transmitter)	VLF (see VLF)
Power, RF	Reception reports 141, 150, 158
ERP, EIRP, difference between 119	RSGB LF Group (see Information sources)
estimating by calculation 118-122	
limits 2, 8, 24, 119	Safety
requirements 77	high antenna voltages . . . 27-28, 41, 49, 97
radiated 2, 23, 121	installing earth stakes 30
Power supplies (see Transmitter)	of your equipment 167, 170
Preamplifier (see Receivers)	protective multiple earthing (PME) 30
Propagation. 129-140	handling ferrite in loading coil 35
aurora 135	when flying kites. 50
coronal mass injection (CME). . . 133, 135	when operating out of doors 153

Selective measuring set
    as receiver 14
    as test gear 14, 116
Software 181-182
    Argo 127, 157, 182
    calculators 181-182
    for SDR 13, 20
    QRS 158, 182
    QRSS programs 182
    Spectrum Laboratory (SpecLab) . . 20, 113,
    . . . . . . 127, 157, 162, 163, 169, 170, 182
Sound card 157, 161, 162, 164
    controls 168
    dongles 170
    external 170
    interfacing with 166
    receiver 170
SWR protection (see also Test equipment)
    . . . . 89-90, 95
Static (see Interference)

T antenna (see Marconi)
Test equipment / measurement . . . .
    . . . . 115-118, 124-127
    antenna tuning meter 126
    dummy load 116
    field strength measurement 122
    frequency counter 116
    inductance meter 116
    oscilloscope 115, 125
    RF current meter 118-119
    RF probe 116
    Scopematch tuning aid 124
    SWR meter 118
Transceiver, all-band
    as receiver (see Receivers)
    as RF source 71
    modifying 71
Transmitter / PA 77-114
    audio amplifiers 79-80, 171
    class D, design notes 80-82
    class D example, 200W, 136kHz . . . 82-87
    class E transmitters 88-89
    EER transverter for LF/MF, an outline . . .
    110-113
    G0MRF 300W, 136kHz 92-99
    G3YXM 1kW, 136kHz 89-92
    G3XBM 10W 472kHz transmit converter
    75
    G4JNT high power 472kHz amplifier . . . .
    104-106
    G4JNT medium power linear for LF/MF
    107-110
    GW3UEP 1W 472kHz 100
    GW3UEP 25W, 472kHz 101
    GW3UEP 100W, 472kHz 102-104
    home made 80-112
    overcurrent protection 95
    power supplies 113
    ready built and kit 77-79
    second-hand and surplus 79
    VLF (see VLF)
Transverter 71-75
    driving by HF transmitter 71
    port on HF transceiver 11

Utility stations (see Non-amateur stations)

VLF 169-172
    antenna 46, 56, 61, 171, 172
    military stations (see non-amateur stations)
    modes 170-171
    permssion to transmit (see Licensing)
    receiving equipment 170
    preselector/amplifier 17-18
    transmitters 171
    what can be heard? 169

Web reporting 51, 148-150, 155
    grabbers 149, 152, 170
    on Opera mode 149
    PSKreporter 150, 160
    remote receivers (see Receivers)
    Reverse Beacon Network 150
    WSPRnet 150, 161
WSPR (see Modes)